Natural Hazards

STEM Road Map
for Elementary School

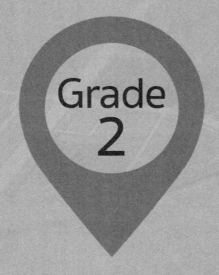

Grade
2

Natural Hazards

Grade 2

STEM Road Map for Elementary School

Edited by Carla C. Johnson, Janet B. Walton, and Erin Peters-Burton

National Science Teaching Association

Arlington, Virginia

National Science Teaching Association

Claire Reinburg, Director
Rachel Ledbetter, Managing Editor
Andrea Silen, Associate Editor
Jennifer Thompson, Associate Editor
Donna Yudkin, Book Acquisitions Manager

ART AND DESIGN
Will Thomas Jr., Director, cover and
  interior design
Himabindu Bichali, Graphic Designer, interior
  design

PRINTING AND PRODUCTION
Catherine Lorrain, Director

NATIONAL SCIENCE TEACHING ASSOCIATION
David L. Evans, Executive Director

1840 Wilson Blvd., Arlington, VA 22201
www.nsta.org/store
For customer service inquiries, please call 800-277-5300.

Copyright © 2019 by the National Science Teaching Association.
All rights reserved. Printed in the United States of America.
22 21 20 19   4 3 2 1

NSTA is committed to publishing material that promotes the best in inquiry-based science education. However, conditions of actual use may vary, and the safety procedures and practices described in this book are intended to serve only as a guide. Additional precautionary measures may be required. NSTA and the authors do not warrant or represent that the procedures and practices in this book meet any safety code or standard of federal, state, or local regulations. NSTA and the authors disclaim any liability for personal injury or damage to property arising out of or relating to the use of this book, including any of the recommendations, instructions, or materials contained therein.

PERMISSIONS
Book purchasers may photocopy, print, or e-mail up to five copies of an NSTA book chapter for personal use only; this does not include display or promotional use. Elementary, middle, and high school teachers may reproduce forms, sample documents, and single NSTA book chapters needed for classroom or noncommercial, professional-development use only. E-book buyers may download files to multiple personal devices but are prohibited from posting the files to third-party servers or websites, or from passing files to non-buyers. For additional permission to photocopy or use material electronically from this NSTA Press book, please contact the Copyright Clearance Center (CCC) (*www.copyright.com*; 978-750-8400). Please access *www.nsta.org/permissions* for further information about NSTA's rights and permissions policies.

Cataloging-in-Publication data for this book and the e-book are available from the Library of Congress.
ISBN: 978-1-68140-486-8
e-ISBN: 978-1-68140-487-5

The *Next Generation Science Standards* ("NGSS") were developed by twenty-six states, in collaboration with the National Research Council, the National Science Teaching Association and the American Association for the Advancement of Science in a process managed by Achieve, Inc. For more information go to *www.nextgenscience.org*.

CONTENTS

CONTENTS

ABOUT THE EDITORS AND AUTHORS

Dr. Carla C. Johnson is executive director of the William and Ida Friday Institute for Educational Innovation, associate dean, and professor of science education in the College of Education at North Carolina State University in Raleigh. She was most recently an associate dean, provost fellow, and professor of science education at Purdue University in West Lafayette, Indiana. Dr. Johnson serves as the director of research and evaluation for the Department of Defense–funded Army Educational Outreach Program (AEOP), a global portfolio of STEM education programs, competitions, and apprenticeships. She has been a leader in STEM education for the past decade, serving as the director of STEM Centers, editor of the *School Science and Mathematics* journal, and lead researcher for the evaluation of Tennessee's Race to the Top–funded STEM portfolio. Dr. Johnson has published over 100 articles, books, book chapters, and curriculum books focused on STEM education. She is a former science and social studies teacher and was the recipient of the 2013 Outstanding Science Teacher Educator of the Year award from the Association for Science Teacher Education (ASTE), the 2012 Award for Excellence in Integrating Science and Mathematics from the School Science and Mathematics Association (SSMA), the 2014 award for best paper on Implications of Research for Educational Practice from ASTE, and the 2006 Outstanding Early Career Scholar Award from SSMA. Her research focuses on STEM education policy implementation, effective science teaching, and integrated STEM approaches.

Dr. Janet B. Walton is a senior research scholar and the assistant director of evaluation for AEOP at North Carolina State University's William and Ida Friday Institute for Educational Innovation. She merges her economic development and education backgrounds to develop K–12 curricular materials that integrate real-life issues with sound cross-curricular content. Her research focuses on mixed methods research methodologies and collaboration between schools and community stakeholders for STEM education and problem- and project-based learning pedagogies. With this research agenda, she works to bring contextual STEM experiences into the classroom and provide students and educators with innovative resources and curricular materials.

Dr. Erin Peters-Burton is the Donna R. and David E. Sterling endowed professor in science education at George Mason University in Fairfax, Virginia. She uses her experiences from 15 years as an engineer and secondary science, engineering, and mathematics teacher to develop research projects that directly inform classroom practice in science and engineering. Her research agenda is based on the idea that all students should build self-awareness of how they learn science and engineering. She works to help students see themselves as "science-minded" and help teachers create classrooms that support student skills to develop scientific knowledge. To accomplish this, she pursues research projects that investigate ways that students and teachers can use self-regulated learning theory in science and engineering, as well as how inclusive STEM schools can help students succeed. During her tenure as a secondary teacher, she had a National Board Certification in Early Adolescent Science and was an Albert Einstein Distinguished Educator Fellow for NASA. As a researcher, Dr. Peters-Burton has published over 100 articles, books, book chapters, and curriculum books focused on STEM education and educational psychology. She received the Outstanding Science Teacher Educator of the Year award from ASTE in 2016 and a Teacher of Distinction Award and a Scholarly Achievement Award from George Mason University in 2012, and in 2010 she was named University Science Educator of the Year by the Virginia Association of Science Teachers.

Dr. Andrea R. Milner is the vice president and dean of academic affairs and an associate professor in the Teacher Education Department at Adrian College in Adrian, Michigan. A former early childhood and elementary teacher, Dr. Milner researches the effects constructivist classroom contextual factors have on student motivation and learning strategy use.

Dr. Tamara J. Moore is an associate professor of engineering education in the College of Engineering at Purdue University. Dr. Moore's research focuses on defining STEM integration through the use of engineering as the connection and investigating its power for student learning.

Dr. Vanessa B. Morrison is an associate professor in the Teacher Education Department at Adrian College. She is a former early childhood teacher and reading and language arts specialist whose research is focused on learning and teaching within a transdisciplinary framework.

Dr. Toni A. Sondergeld is an associate professor of assessment, research, and statistics in the School of Education at Drexel University in Philadelphia. Dr. Sondergeld's research concentrates on assessment and evaluation in education, with a focus on K–12 STEM.

ACKNOWLEDGMENTS

This module was developed as a part of the STEM Road Map project (Carla C. Johnson, principal investigator). The Purdue University College of Education, General Motors, and other sources provided funding for this project.

Copyright © 2015 from *STEM Road Map: A Framework for Integrated STEM Education,* edited by C. C. Johnson, E. E. Peters-Burton, and T. J. Moore. Reproduced by permission of Taylor and Francis Group, LLC, a division of Informa plc.

See *www.routledge.com/products/9781138804234* for more information about *STEM Road Map: A Framework for Integrated STEM Education.*

PART 1

THE STEM ROAD MAP

BACKGROUND, THEORY, AND PRACTICE

OVERVIEW OF THE *STEM ROAD MAP CURRICULUM SERIES*

Carla C. Johnson, Erin Peters-Burton, and Tamara J. Moore

The *STEM Road Map Curriculum Series* was conceptualized and developed by a team of STEM educators from across the United States in response to a growing need to infuse real-world learning contexts, delivered through authentic problem-solving pedagogy, into K–12 classrooms. The curriculum series is grounded in integrated STEM, which focuses on the integration of the STEM disciplines—science, technology, engineering, and mathematics—delivered across content areas, incorporating the Framework for 21st Century Learning along with grade-level-appropriate academic standards.

The curriculum series begins in kindergarten, with a five-week instructional sequence that introduces students to the STEM themes and gives them grade-level-appropriate topics and real-world challenges or problems to solve. The series uses project-based and problem-based learning, presenting students with the problem or challenge during the first lesson, and then teaching them science, social studies, English language arts, mathematics, and other content, as they apply what they learn to the challenge or problem at hand.

Authentic assessment and differentiation are embedded throughout the modules. Each *STEM Road Map Curriculum Series* module has a lead discipline, which may be science, social studies, English language arts, or mathematics. All disciplines are integrated into each module, along with ties to engineering. Another key component is the use of STEM Research Notebooks to allow students to track their own learning progress. The modules are designed with a scaffolded approach, with increasingly complex concepts and skills introduced as students progress through grade levels.

The developers of this work view the curriculum as a resource that is intended to be used either as a whole or in part to meet the needs of districts, schools, and teachers who are implementing an integrated STEM approach. A variety of implementation formats are possible, from using one stand-alone module at a given grade level to using all five modules to provide 25 weeks of instruction. Also, within each grade band (K–2, 3–5, 6–8, 9–12), the modules can be sequenced in various ways to suit specific needs.

STANDARDS-BASED APPROACH

The *STEM Road Map Curriculum Series* is anchored in the *Next Generation Science Standards* (*NGSS*), the *Common Core State Standards for Mathematics* (*CCSS Mathematics*), the *Common Core State Standards for English Language Arts* (*CCSS ELA*), and the Framework for 21st Century Learning. Each module includes a detailed curriculum map that incorporates the associated standards from the particular area correlated to lesson plans. The STEM Road Map has very clear and strong connections to these academic standards, and each of the grade-level topics was derived from the mapping of the standards to ensure alignment among topics, challenges or problems, and the required academic standards for students. Therefore, the curriculum series takes a standards-based approach and is designed to provide authentic contexts for application of required knowledge and skills.

THEMES IN THE *STEM ROAD MAP CURRICULUM SERIES*

The K–12 STEM Road Map is organized around five real-world STEM themes that were generated through an examination of the big ideas and challenges for society included in STEM standards and those that are persistent dilemmas for current and future generations:

- Cause and Effect
- Innovation and Progress
- The Represented World
- Sustainable Systems
- Optimizing the Human Experience

These themes are designed as springboards for launching students into an exploration of real-world learning situated within big ideas. Most important, the five STEM Road Map themes serve as a framework for scaffolding STEM learning across the K–12 continuum.

The themes are distributed across the STEM disciplines so that they represent the big ideas in science (Cause and Effect; Sustainable Systems), technology (Innovation and Progress; Optimizing the Human Experience), engineering (Innovation and Progress; Sustainable Systems; Optimizing the Human Experience), and mathematics (The Represented World), as well as concepts and challenges in social studies and 21st century skills that are also excellent contexts for learning in English language arts. The process of developing themes began with the clustering of the *NGSS* performance expectations and the National Academy of Engineering's grand challenges for engineering, which led to the development of the challenge in each module and connections of the module activities to the *CCSS Mathematics* and *CCSS ELA* standards. We performed these

mapping processes with large teams of experts and found that these five themes provided breadth, depth, and coherence to frame a high-quality STEM learning experience from kindergarten through 12th grade.

Cause and Effect

The concept of cause and effect is a powerful and pervasive notion in the STEM fields. It is the foundation of understanding how and why things happen as they do. Humans spend considerable effort and resources trying to understand the causes and effects of natural and designed phenomena to gain better control over events and the environment and to be prepared to react appropriately. Equipped with the knowledge of a specific cause-and-effect relationship, we can lead better lives or contribute to the community by altering the cause, leading to a different effect. For example, if a person recognizes that irresponsible energy consumption leads to global climate change, that person can act to remedy his or her contribution to the situation. Although cause and effect is a core idea in the STEM fields, it can actually be difficult to determine. Students should be capable of understanding not only when evidence points to cause and effect but also when evidence points to relationships but not direct causality. The major goal of education is to foster students to be empowered, analytic thinkers, capable of thinking through complex processes to make important decisions. Understanding causality, as well as when it cannot be determined, will help students become better consumers, global citizens, and community members.

Innovation and Progress

One of the most important factors in determining whether humans will have a positive future is innovation. Innovation is the driving force behind progress, which helps create possibilities that did not exist before. Innovation and progress are creative entities, but in the STEM fields, they are anchored by evidence and logic, and they use established concepts to move the STEM fields forward. In creating something new, students must consider what is already known in the STEM fields and apply this knowledge appropriately. When we innovate, we create value that was not there previously and create new conditions and possibilities for even more innovations. Students should consider how their innovations might affect progress and use their STEM thinking to change current human burdens to benefits. For example, if we develop more efficient cars that use by-products from another manufacturing industry, such as food processing, then we have used waste productively and reduced the need for the waste to be hauled away, an indirect benefit of the innovation.

The Represented World

When we communicate about the world we live in, how the world works, and how we can meet the needs of humans, sometimes we can use the actual phenomena to explain a concept. Sometimes, however, the concept is too big, too slow, too small, too fast, or too complex for us to explain using the actual phenomena, and we must use a representation or a model to help communicate the important features. We need representations and models such as graphs, tables, mathematical expressions, and diagrams because it makes our thinking visible. For example, when examining geologic time, we cannot actually observe the passage of such large chunks of time, so we create a timeline or a model that uses a proportional scale to visually illustrate how much time has passed for different eras. Another example may be something too complex for students at a particular grade level, such as explaining the *p* subshell orbitals of electrons to fifth graders. Instead, we use the Bohr model, which more closely represents the orbiting of planets and is accessible to fifth graders.

When we create models, they are helpful because they point out the most important features of a phenomenon. We also create representations of the world with mathematical functions, which help us change parameters to suit the situation. Creating representations of a phenomenon engages students because they are able to identify the important features of that phenomenon and communicate them directly. But because models are estimates of a phenomenon, they leave out some of the details, so it is important for students to evaluate their usefulness as well as their shortcomings.

Sustainable Systems

From an engineering perspective, the term *system* refers to the use of "concepts of component need, component interaction, systems interaction, and feedback. The interaction of subcomponents to produce a functional system is a common lens used by all engineering disciplines for understanding, analysis, and design." (Koehler, Bloom, and Binns 2013, p. 8). Systems can be either open (e.g., an ecosystem) or closed (e.g., a car battery). Ideally, a system should be sustainable, able to maintain equilibrium without much energy from outside the structure. Looking at a garden, we see flowers blooming, weeds sprouting, insects buzzing, and various forms of life living within its boundaries. This is an example of an ecosystem, a collection of living organisms that survive together, functioning as a system. The interaction of the organisms within the system and the influences of the environment (e.g., water, sunlight) can maintain the system for a period of time, thus demonstrating its ability to endure. Sustainability is a desirable feature of a system because it allows for existence of the entity in the long term.

In the STEM Road Map project, we identified different standards that we consider to be oriented toward systems that students should know and understand in the K–12 setting. These include ecosystems, the rock cycle, Earth processes (such as erosion, tectonics, ocean currents, weather phenomena), Earth-Sun-Moon cycles, heat transfer, and the interaction among the geosphere, biosphere, hydrosphere, and atmosphere. Students and teachers should understand that we live in a world of systems that are not independent of each other, but rather are intrinsically linked such that a disruption in one part of a system will have reverberating effects on other parts of the system.

Optimizing the Human Experience

Science, technology, engineering, and mathematics as disciplines have the capacity to continuously improve the ways humans live, interact, and find meaning in the world, thus working to optimize the human experience. This idea has two components: being more suited to our environment and being more fully human. For example, the progression of STEM ideas can help humans create solutions to complex problems, such as improving ways to access water sources, designing energy sources with minimal impact on our environment, developing new ways of communication and expression, and building efficient shelters. STEM ideas can also provide access to the secrets and wonders of nature. Learning in STEM requires students to think logically and systematically, which is a way of knowing the world that is markedly different from knowing the world as an artist. When students can employ various ways of knowing and understand when it is appropriate to use a different way of knowing or integrate ways of knowing, they are fully experiencing the best of what it is to be human. The problem-based learning scenarios provided in the STEM Road Map help students develop ways of thinking like STEM professionals as they ask questions and design solutions. They learn to optimize the human experience by innovating improvements in the designed world in which they live.

THE NEED FOR AN INTEGRATED STEM APPROACH

At a basic level, STEM stands for science, technology, engineering, and mathematics. Over the past decade, however, STEM has evolved to have a much broader scope and broader implications. Now, educators and policy makers refer to STEM as not only a concentrated area for investing in the future of the United States and other nations but also as a domain and mechanism for educational reform.

The good intentions of the recent decade-plus of focus on accountability and increased testing has resulted in significant decreases not only in instructional time for teaching science and social studies but also in the flexibility of teachers to promote authentic, problem solving–focused classroom environments. The shift has had a detrimental impact on student acquisition of vitally important skills, which many refer to as 21st century skills, and often the ability of students to "think." Further, schooling has become increasingly siloed into compartments of mathematics, science, English language arts, and social studies, lacking any of the connections that are overwhelmingly present in

the real world around children. Students have experienced school as content provided in boxes that must be memorized, devoid of any real-world context, and often have little understanding of why they are learning these things.

STEM-focused projects, curriculum, activities, and schools have emerged as a means to address these challenges. However, most of these efforts have continued to focus on the individual STEM disciplines (predominantly science and engineering) through more STEM classes and after-school programs in a "STEM enhanced" approach (Breiner et al. 2012). But in traditional and STEM enhanced approaches, there is little to no focus on other disciplines that are integral to the context of STEM in the real world. Integrated STEM education, on the other hand, infuses the learning of important STEM content and concepts with a much-needed emphasis on 21st century skills and a problem- and project-based pedagogy that more closely mirrors the real-world setting for society's challenges. It incorporates social studies, English language arts, and the arts as pivotal and necessary (Johnson 2013; Rennie, Venville, and Wallace 2012; Roehrig et al. 2012).

FRAMEWORK FOR STEM INTEGRATION IN THE CLASSROOM

The *STEM Road Map Curriculum Series* is grounded in the Framework for STEM Integration in the Classroom as conceptualized by Moore, Guzey, and Brown (2014) and Moore et al. (2014). The framework has six elements, described in the context of how they are used in the *STEM Road Map Curriculum Series* as follows:

1. The STEM Road Map contexts are meaningful to students and provide motivation to engage with the content. Together, these allow students to have different ways to enter into the challenge.

2. The STEM Road Map modules include engineering design that allows students to design technologies (i.e., products that are part of the designed world) for a compelling purpose.

3. The STEM Road Map modules provide students with the opportunities to learn from failure and redesign based on the lessons learned.

4. The STEM Road Map modules include standards-based disciplinary content as the learning objectives.

5. The STEM Road Map modules include student-centered pedagogies that allow students to grapple with the content, tie their ideas to the context, and learn to think for themselves as they deepen their conceptual knowledge.

6. The STEM Road Map modules emphasize 21st century skills and, in particular, highlight communication and teamwork.

All of the STEM Road Map modules incorporate these six elements; however, the level of emphasis on each of these elements varies based on the challenge or problem in each module.

THE NEED FOR THE *STEM ROAD MAP CURRICULUM SERIES*

As focus is increasing on integrated STEM, and additional schools and programs decide to move their curriculum and instruction in this direction, there is a need for high-quality, research-based curriculum designed with integrated STEM at the core. Several good resources are available to help teachers infuse engineering or more STEM enhanced approaches, but no curriculum exists that spans K–12 with an integrated STEM focus. The next chapter provides detailed information about the specific pedagogy, instructional strategies, and learning theory on which the *STEM Road Map Curriculum Series* is grounded.

REFERENCES

Breiner, J., M. Harkness, C. C. Johnson, and C. Koehler. 2012. What is STEM? A discussion about conceptions of STEM in education and partnerships. *School Science and Mathematics* 112 (1): 3–11.

Johnson, C. C. 2013. Conceptualizing integrated STEM education: Editorial. *School Science and Mathematics* 113 (8): 367–368.

Koehler, C. M., M. A. Bloom, and I. C. Binns. 2013. Lights, camera, action: Developing a methodology to document mainstream films' portrayal of nature of science and scientific inquiry. *Electronic Journal of Science Education* 17 (2).

Moore, T. J., S. S. Guzey, and A. Brown. 2014. Greenhouse design to increase habitable land: An engineering unit. *Science Scope* 37 (7): 51–57.

Moore, T. J., M. S. Stohlmann, H.-H. Wang, K. M. Tank, A. W. Glancy, and G. H. Roehrig. 2014. Implementation and integration of engineering in K–12 STEM education. In *Engineering in pre-college settings: Synthesizing research, policy, and practices,* ed. S. Purzer, J. Strobel, and M. Cardella, 35–60. West Lafayette, IN: Purdue Press.

Rennie, L., G. Venville, and J. Wallace. 2012. *Integrating science, technology, engineering, and mathematics: Issues, reflections, and ways forward.* New York: Routledge.

Roehrig, G. H., T. J. Moore, H. H. Wang, and M. S. Park. 2012. Is adding the E enough? Investigating the impact of K–12 engineering standards on the implementation of STEM integration. *School Science and Mathematics* 112 (1): 31–44.

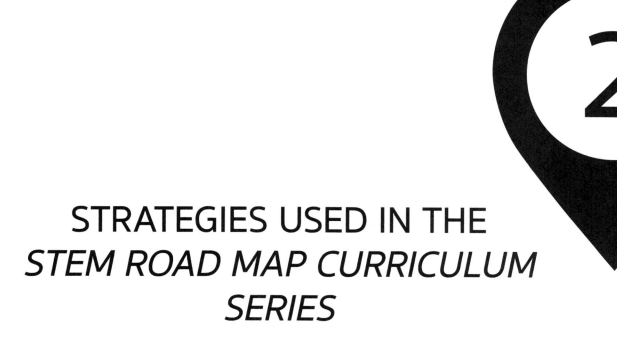

STRATEGIES USED IN THE *STEM ROAD MAP CURRICULUM SERIES*

Erin Peters-Burton, Carla C. Johnson, Toni A. Sondergeld, and Tamara J. Moore

The *STEM Road Map Curriculum Series* uses what has been identified through research as best-practice pedagogy, including embedded formative assessment strategies throughout each module. This chapter briefly describes the key strategies that are employed in the series.

PROJECT- AND PROBLEM-BASED LEARNING

Each module in the *STEM Road Map Curriculum Series* uses either project-based learning or problem-based learning to drive the instruction. Project-based learning begins with a driving question to guide student teams in addressing a contextualized local or community problem or issue. The outcome of project-based instruction is a product that is conceptualized, designed, and tested through a series of scaffolded learning experiences (Blumenfeld et al. 1991; Krajcik and Blumenfeld 2006). Problem-based learning is often grounded in a fictitious scenario, challenge, or problem (Barell 2006; Lambros 2004). On the first day of instruction within the unit, student teams are provided with the context of the problem. Teams work through a series of activities and use open-ended research to develop their potential solution to the problem or challenge, which need not be a tangible product (Johnson 2003).

ENGINEERING DESIGN PROCESS

The *STEM Road Map Curriculum Series* uses engineering design as a way to facilitate integrated STEM within the modules. The engineering design process (EDP) is depicted in Figure 2.1 (p. 10). It highlights two major aspects of engineering design—problem scoping and solution generation—and six specific components of working toward a design: define the problem, learn about the problem, plan a solution, try the solution, test the solution, decide whether the solution is good enough. It also shows that communication

Figure 2.1. Engineering Design Process

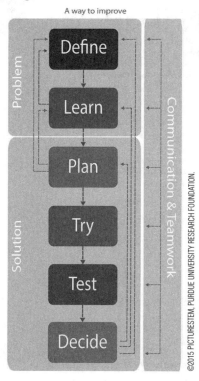

and teamwork are involved throughout the entire process. As the arrows in the figure indicate, the order in which the components of engineering design are addressed depends on what becomes needed as designers progress through the EDP. Designers must communicate and work in teams throughout the process. The EDP is iterative, meaning that components of the process can be repeated as needed until the design is good enough to present to the client as a potential solution to the problem.

Problem scoping is the process of gathering and analyzing information to deeply understand the engineering design problem. It includes defining the problem and learning about the problem. Defining the problem includes identifying the problem, the client, and the end user of the design. The client is the person (or people) who hired the designers to do the work, and the end user is the person (or people) who will use the final design. The designers must also identify the criteria and the constraints of the problem. The criteria are the things the client wants from the solution, and the constraints are the things that limit the possible solutions. The designers must spend significant time learning about the problem, which can include activities such as the following:

- Reading informational texts and researching about relevant concepts or contexts

- Identifying and learning about needed mathematical and scientific skills, knowledge, and tools

- Learning about things done previously to solve similar problems

- Experimenting with possible materials that could be used in the design

Problem scoping also allows designers to consider how to measure the success of the design in addressing specific criteria and staying within the constraints over multiple iterations of solution generation.

Solution generation includes planning a solution, trying the solution, testing the solution, and deciding whether the solution is good enough. Planning the solution includes generating many design ideas that both address the criteria and meet the constraints. Here the designers must consider what was learned about the problem during problem scoping. Design plans include clear communication of design ideas through media such as notebooks, blueprints, schematics, or storyboards. They also include details about the

design, such as measurements, materials, colors, costs of materials, instructions for how things fit together, and sets of directions. Making the decision about which design idea to move forward involves considering the trade-offs of each design idea.

Once a clear design plan is in place, the designers must try the solution. Trying the solution includes developing a prototype (a testable model) based on the plan generated. The prototype might be something physical or a process to accomplish a goal. This component of design requires that the designers consider the risk involved in implementing the design. The prototype developed must be tested. Testing the solution includes conducting fair tests that verify whether the plan is a solution that is good enough to meet the client and end user needs and wants. Data need to be collected about the results of the tests of the prototype, and these data should be used to make evidence-based decisions regarding the design choices made in the plan. Here, the designers must again consider the criteria and constraints for the problem.

Using the data gathered from the testing, the designers must decide whether the solution is good enough to meet the client and end user needs and wants by assessment based on the criteria and constraints. Here, the designers must justify or reject design decisions based on the background research gathered while learning about the problem and on the evidence gathered during the testing of the solution. The designers must now decide whether to present the current solution to the client as a possibility or to do more iterations of design on the solution. If they decide that improvements need to be made to the solution, the designers must decide if there is more that needs to be understood about the problem, client, or end user; if another design idea should be tried; or if more planning needs to be conducted on the same design. One way or another, more work needs to be done.

Throughout the process of designing a solution to meet a client's needs and wants, designers work in teams and must communicate to each other, the client, and likely the end user. Teamwork is important in engineering design because multiple perspectives and differing skills and knowledge are valuable when working to solve problems. Communication is key to the success of the designed solution. Designers must communicate their ideas clearly using many different representations, such as text in an engineering notebook, diagrams, flowcharts, technical briefs, or memos to the client.

LEARNING CYCLE

The same format for the learning cycle is used in all grade levels throughout the STEM Road Map, so that students engage in a variety of activities to learn about phenomena in the modules thoroughly and have consistent experiences in the problem- and project-based learning modules. Expectations for learning by younger students are not as high as for older students, but the format of the progression of learning is the same. Students who have learned with curriculum from the STEM Road Map in early grades know

what to expect in later grades. The learning cycle consists of five parts—Introductory Activity/Engagement, Activity/Exploration, Explanation, Elaboration/Application of Knowledge, and Evaluation/Assessment—and is based on the empirically tested 5E model from BSCS (Bybee et al. 2006).

In the Introductory Activity/Engagement phase, teachers introduce the module challenge and use a unique approach designed to pique students' curiosity. This phase gets students to start thinking about what they already know about the topic and begin wondering about key ideas. The Introductory Activity/Engagement phase positions students to be confident about what they are about to learn, because they have prior knowledge, and clues them into what they don't yet know.

In the Activity/Exploration phase, the teacher sets up activities in which students experience a deeper look at the topics that were introduced earlier. Students engage in the activities and generate new questions or consider possibilities using preliminary investigations. Students work independently, in small groups, and in whole-group settings to conduct investigations, resulting in common experiences about the topic and skills involved in the real-world activities. Teachers can assess students' development of concepts and skills based on the common experiences during this phase.

During the Explanation phase, teachers direct students' attention to concepts they need to understand and skills they need to possess to accomplish the challenge. Students participate in activities to demonstrate their knowledge and skills to this point, and teachers can pinpoint gaps in student knowledge during this phase.

In the Elaboration/Application of Knowledge phase, teachers present students with activities that engage in higher-order thinking to create depth and breadth of student knowledge, while connecting ideas across topics within and across STEM. Students apply what they have learned thus far in the module to a new context or elaborate on what they have learned about the topic to a deeper level of detail.

In the last phase, Evaluation/Assessment, teachers give students summative feedback on their knowledge and skills as demonstrated through the challenge. This is not the only point of assessment (as discussed in the section on Embedded Formative Assessments), but it is an assessment of the culmination of the knowledge and skills for the module. Students demonstrate their cognitive growth at this point and reflect on how far they have come since the beginning of the module. The challenges are designed to be multidimensional in the ways students must collaborate and communicate their new knowledge.

STEM RESEARCH NOTEBOOK

One of the main components of the *STEM Road Map Curriculum Series* is the STEM Research Notebook, a place for students to capture their ideas, questions, observations, reflections, evidence of progress, and other items associated with their daily work. At the beginning of each module, the teacher walks students through the setup of the STEM

Research Notebook, which could be a three-ring binder, composition book, or spiral notebook. You may wish to have students create divided sections so that they can easily access work from various disciplines during the module. Electronic notebooks kept on student devices are also acceptable and encouraged. Students will develop their own table of contents and create chapters in the notebook for each module.

Each lesson in the *STEM Road Map Curriculum Series* includes one or more prompts that are designed for inclusion in the STEM Research Notebook and appear as questions or statements that the teacher assigns to students. These prompts require students to apply what they have learned across the lesson to solve the big problem or challenge for that module. Each lesson is designed to meaningfully refer students to the larger problem or challenge they have been assigned to solve with their teams. The STEM Research Notebook is designed to be a key formative assessment tool, as students' daily entries provide evidence of what they are learning. The notebook can be used as a mechanism for dialogue between the teacher and students, as well as for peer and self-evaluation.

The use of the STEM Research Notebook is designed to scaffold student notebooking skills across the grade bands in the *STEM Road Map Curriculum Series*. In the early grades, children learn how to organize their daily work in the notebook as a way to collect their products for future reference. In elementary school, students structure their notebooks to integrate background research along with their daily work and lesson prompts. In the upper grades (middle and high school), students expand their use of research and data gathering through team discussions to more closely mirror the work of STEM experts in the real world.

THE ROLE OF ASSESSMENT IN THE *STEM ROAD MAP CURRICULUM SERIES*

Starting in the middle years and continuing into secondary education, the word *assessment* typically brings grades to mind. These grades may take the form of a letter or a percentage, but they typically are used as a representation of a student's content mastery. If well thought out and implemented, however, classroom assessment can offer teachers, parents, and students valuable information about student learning and misconceptions that does not necessarily come in the form of a grade (Popham 2013).

The *STEM Road Map Curriculum Series* provides a set of assessments for each module. Teachers are encouraged to use assessment information for more than just assigning grades to students. Instead, assessments of activities requiring students to actively engage in their learning, such as student journaling in STEM Research Notebooks, collaborative presentations, and constructing graphic organizers, should be used to move student learning forward. Whereas other curriculum with assessments may include objective-type (multiple-choice or matching) tests, quizzes, or worksheets, we have intentionally avoided these forms of assessments to better align assessment strategies with teacher instruction and

student learning techniques. Since the focus of this book is on project- or problem-based STEM curriculum and instruction that focuses on higher-level thinking skills, appropriate and authentic performance assessments were developed to elicit the most reliable and valid indication of growth in student abilities (Brookhart and Nitko 2008).

Comprehensive Assessment System

Assessment throughout all STEM Road Map curriculum modules acts as a comprehensive system in which formative and summative assessments work together to provide teachers with high-quality information on student learning. Formative assessment occurs when the teacher finds out formally or informally what a student knows about a smaller, defined concept or skill and provides timely feedback to the student about his or her level of proficiency. Summative assessments occur when students have performed all activities in the module and are given a cumulative performance evaluation in which they demonstrate their growth in learning.

A comprehensive assessment system can be thought of as akin to a sporting event. Formative assessments are the practices: It is important to accomplish them consistently, they provide feedback to help students improve their learning, and making mistakes can be worthwhile if students are given an opportunity to learn from them. Summative assessments are the competitions: Students need to be prepared to perform at the best of their ability. Without multiple opportunities to practice skills along the way through formative assessments, students will not have the best chance of demonstrating growth in abilities through summative assessments (Black and Wiliam 1998).

Embedded Formative Assessments

Formative assessments in this module serve two main purposes: to provide feedback to students about their learning and to provide important information for the teacher to inform immediate instructional needs. Providing feedback to students is particularly important when conducting problem- or project-based learning because students take on much of the responsibility for learning, and teachers must facilitate student learning in an informed way. For example, if students are required to conduct research for the Activity/Exploration phase but are not familiar with what constitutes a reliable resource, they may develop misconceptions based on poor information. When a teacher monitors this learning through formative assessments and provides specific feedback related to the instructional goals, students are less likely to develop incomplete or incorrect conceptions in their independent investigations. By using formative assessment to detect problems in student learning and then acting on this information, teachers help move student learning forward through these teachable moments.

Formative assessments come in a variety of formats. They can be informal, such as asking students probing questions related to student knowledge or tasks or simply observing students engaged in an activity to gather information about student skills. Formative assessments can also be formal, such as a written quiz or a laboratory practical. Regardless of the type, three key steps must be completed when using formative assessments (Sondergeld, Bell, and Leusner 2010). First, the assessment is delivered to students so that teachers can collect data. Next, teachers analyze the data (student responses) to determine student strengths and areas that need additional support. Finally, teachers use the results from information collected to modify lessons and create learning environments that reinforce weak points in student learning. If student learning information is not used to modify instruction, the assessment cannot be considered formative in nature.

Formative assessments can be about content, science process skills, or even learning skills. When a formative assessment focuses on content, it assesses student knowledge about the disciplinary core ideas from the *Next Generation Science Standards* (*NGSS*) or content objectives from *Common Core State Standards for Mathematics* (*CCSS Mathematics*) or *Common Core State Standards for English Language Arts* (*CCSS ELA*). Content-focused formative assessments ask students questions about declarative knowledge regarding the concepts they have been learning. Process skills formative assessments examine the extent to which a student can perform science and engineering practices from the *NGSS* or process objectives from *CCSS Mathematics* or *CCSS ELA*, such as constructing an argument. Learning skills can also be assessed formatively by asking students to reflect on the ways they learn best during a module and identify ways they could have learned more.

Assessment Maps

Assessment maps or blueprints can be used to ensure alignment between classroom instruction and assessment. If what students are learning in the classroom is not the same as the content on which they are assessed, the resultant judgment made on student learning will be invalid (Brookhart and Nitko 2008). Therefore, the issue of instruction and assessment alignment is critical. The assessment map for this book (found in Chapter 3) indicates by lesson whether the assessment should be completed as a group or on an individual basis, identifies the assessment as formative or summative in nature, and aligns the assessment with its corresponding learning objectives.

Note that the module includes far more formative assessments than summative assessments. This is done intentionally to provide students with multiple opportunities to practice their learning of new skills before completing a summative assessment. Note also that formative assessments are used to collect information on only one or two learning objectives at a time so that potential relearning or instructional modifications can focus on smaller and more manageable chunks of information. Conversely, summative assessments in the module cover many more learning objectives, as they are traditionally used as final markers of student learning. This is not to say that information collected from summative assessments cannot or should not be used formatively. If teachers find that gaps in student

learning persist after a summative assessment is completed, it is important to revisit these existing misconceptions or areas of weakness before moving on (Black et al. 2003).

SELF-REGULATED LEARNING THEORY IN THE STEM ROAD MAP MODULES

Many learning theories are compatible with the STEM Road Map modules, such as constructivism, situated cognition, and meaningful learning. However, we feel that the self-regulated learning theory (SRL) aligns most appropriately (Zimmerman 2000). SRL requires students to understand that thinking needs to be motivated and managed (Ritchhart, Church, and Morrison 2011). The STEM Road Map modules are student centered and are designed to provide students with choices, concrete hands-on experiences, and opportunities to see and make connections, especially across subjects (Eliason and Jenkins 2012; NAEYC 2016). Additionally, SRL is compatible with the modules because it fosters a learning environment that supports students' motivation, enables students to become aware of their own learning strategies, and requires reflection on learning while experiencing the module (Peters and Kitsantas 2010).

The theory behind SRL (see Figure 2.2) explains the different processes that students engage in before, during, and after a learning task. Because SRL is a cyclical learning process, the accomplishment of one cycle develops strategies for the next learning cycle. This cyclic way of learning aligns with the various sections in the STEM Road Map lesson plans on Introductory Activity/Engagement, Activity/Exploration, Explanation, Elaboration/Application of Knowledge, and Evaluation/Assessment. Since the students engaged in a module take on much of the responsibility for learning, this theory also provides guidance for teachers to keep students on the right track.

The remainder of this section explains how SRL theory is embedded within the five sections of each module and points out ways to

Figure 2.2. SRL Theory

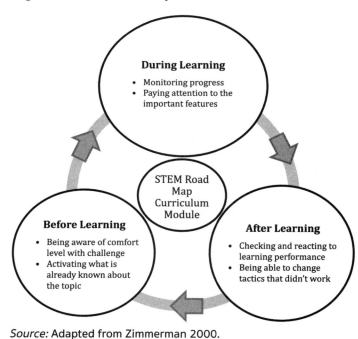

Source: Adapted from Zimmerman 2000.

support students in becoming independent learners of STEM while productively functioning in collaborative teams.

Before Learning: Setting the Stage

Before attempting a learning task such as the STEM Road Map modules, teachers should develop an understanding of their students' level of comfort with the process of accomplishing the learning and determine what they already know about the topic. When students are comfortable with attempting a learning task, they tend to take more risks in learning and as a result achieve deeper learning (Bandura 1986).

The STEM Road Map curriculum modules are designed to foster excitement from the very beginning. Each module has an Introductory Activity/Engagement section that introduces the overall topic from a unique and exciting perspective, engaging the students to learn more so that they can accomplish the challenge. The Introductory Activity also has a design component that helps teachers assess what students already know about the topic of the module. In addition to the deliberate designs in the lesson plans to support SRL, teachers can support a high level of student comfort with the learning challenge by finding out if students have ever accomplished the same kind of task and, if so, asking them to share what worked well for them.

During Learning: Staying the Course

Some students fear inquiry learning because they aren't sure what to do to be successful (Peters 2010). However, the STEM Road Map curriculum modules are embedded with tools to help students pay attention to knowledge and skills that are important for the learning task and to check student understanding along the way. One of the most important processes for learning is the ability for learners to monitor their own progress while performing a learning task (Peters 2012). The modules allow students to monitor their progress with tools such as the STEM Research Notebooks, in which they record what they know and can check whether they have acquired a complete set of knowledge and skills. The STEM Road Map modules support inquiry strategies that include previewing, questioning, predicting, clarifying, observing, discussing, and journaling (Morrison and Milner 2014). Through the use of technology throughout the modules, inquiry is supported by providing students access to resources and data while enabling them to process information, report the findings, collaborate, and develop 21st century skills.

It is important for teachers to encourage students to have an open mind about alternative solutions and procedures (Milner and Sondergeld 2015) when working through the STEM Road Map curriculum modules. Novice learners can have difficulty knowing what to pay attention to and tend to treat each possible avenue for information as equal (Benner 1984). Teachers are the mentors in a classroom and can point out ways for students to approach learning during the Activity/Exploration, Explanation, and

Elaboration/Application of Knowledge portions of the lesson plans to ensure that students pay attention to the important concepts and skills throughout the module. For example, if a student is to demonstrate conceptual awareness of motion when working on roller coaster research, but the student has misconceptions about motion, the teacher can step in and redirect student learning.

After Learning: Knowing What Works

The classroom is a busy place, and it may often seem that there is no time for self-reflection on learning. Although skipping this reflective process may save time in the short term, it reduces the ability to take into account things that worked well and things that didn't so that teaching the module may be improved next time. In the long run, SRL skills are critical for students to become independent learners who can adapt to new situations. By investing the time it takes to teach students SRL skills, teachers can save time later, because students will be able to apply methods and approaches for learning that they have found effective to new situations. In the Evaluation/Assessment portion of the STEM Road Map curriculum modules, as well as in the formative assessments throughout the modules, two processes in the after-learning phase are supported: evaluating one's own performance and accounting for ways to adapt tactics that didn't work well. Students have many opportunities to self-assess in formative assessments, both in groups and individually, using the rubrics provided in the modules.

The designs of the *NGSS* and *CCSS* allow for students to learn in diverse ways, and the STEM Road Map curriculum modules emphasize that students can use a variety of tactics to complete the learning process. For example, students can use STEM Research Notebooks to record what they have learned during the various research activities. Notebook entries might include putting objectives in students' own words, compiling their prior learning on the topic, documenting new learning, providing proof of what they learned, and reflecting on what they felt successful doing and what they felt they still needed to work on. Perhaps students didn't realize that they were supposed to connect what they already knew with what they learned. They could record this and would be prepared in the next learning task to begin connecting prior learning with new learning.

SAFETY IN STEM

Student safety is a primary consideration in all subjects but is an area of particular concern in science, where students may interact with unfamiliar tools and materials that may pose additional safety risks. It is important to implement safety practices within the context of STEM investigations, whether in a classroom laboratory or in the field. When you keep safety in mind as a teacher, you avoid many potential issues with the lesson while also protecting your students.

STEM safety practices encompass things considered in the typical science classroom. Ensure that students are familiar with basic safety considerations, such as wearing protective equipment (e.g., safety glasses or goggles and latex-free gloves) and taking care with sharp objects, and know emergency exit procedures. Teachers should learn beforehand the locations of the safety eyewash, fume hood, fire extinguishers, and emergency shut-off switch in the classroom and how to use them. Also be aware of any school or district safety policies that are in place and apply those that align with the work being conducted in the lesson. It is important to review all safety procedures annually.

STEM investigations should always be supervised. Each lesson in the modules includes teacher guidelines for applicable safety procedures that should be followed. Before each investigation, teachers should go over these safety procedures with the student teams. Some STEM focus areas such as engineering require that students can demonstrate how to properly use equipment in the maker space before the teacher allows them to proceed with the lesson.

Information about classroom science safety, including a safety checklist for science classrooms, general lab safety recommendations, and links to other science safety resources, is available at the Council of State Science Supervisors (CSSS) website at *www.csss-science. org/safety.shtml*. The National Science Teaching Association (NSTA) provides a list of science rules and regulations, including standard operating procedures for lab safety, and a safety acknowledgment form for students and parents or guardians to sign. You can access these resources at *http://static.nsta.org/pdfs/SafetyInTheScienceClassroom.pdf*. In addition, NSTA's Safety in the Science Classroom web page (*www.nsta.org/safety*) has numerous links to safety resources, including papers written by the NSTA Safety Advisory Board.

Disclaimer: The safety precautions for each activity are based on use of the recommended materials and instructions, legal safety standards, and better professional practices. Using alternative materials or procedures for these activities may jeopardize the level of safety and therefore is at the user's own risk.

REFERENCES

Bandura, A. 1986. *Social foundations of thought and action: A social cognitive theory.* Englewood Cliffs, NJ: Prentice-Hall.

Barell, J. 2006. *Problem-based learning: An inquiry approach.* Thousand Oaks, CA: Corwin Press.

Benner, P. 1984. *From novice to expert: Excellence and power in clinical nursing practice.* Menlo Park, CA: Addison-Wesley.

Black, P., C. Harrison, C. Lee, B. Marshall, and D. Wiliam. 2003. *Assessment for learning: Putting it into practice.* Berkshire, UK: Open University Press.

Black, P., and D. Wiliam. 1998. Inside the black box: Raising standards through classroom assessment. *Phi Delta Kappan* 80 (2): 139–148.

Blumenfeld, P., E. Soloway, R. Marx, J. Krajcik, M. Guzdial, and A. Palincsar. 1991. Motivating project-based learning: Sustaining the doing, supporting learning. *Educational Psychologist* 26 (3): 369–398.

Brookhart, S. M., and A. J. Nitko. 2008. *Assessment and grading in classrooms.* Upper Saddle River, NJ: Pearson.

Bybee, R., J. Taylor, A. Gardner, P. Van Scotter, J. Carlson Powell, A. Westbrook, and N. Landes. 2006. *The BSCS 5E instructional model: Origins and effectiveness.* Colorado Springs, CO: BSCS.

Eliason, C. F., and L. T. Jenkins. 2012. *A practical guide to early childhood curriculum.* 9th ed. New York: Merrill.

Johnson, C. 2003. Bioterrorism is real-world science: Inquiry-based simulation mirrors real life. *Science Scope* 27 (3): 19–23.

Krajcik, J., and P. Blumenfeld. 2006. Project-based learning. In *The Cambridge handbook of the learning sciences,* ed. R. Keith Sawyer, 317–334. New York: Cambridge University Press.

Lambros, A. 2004. *Problem-based learning in middle and high school classrooms: A teacher's guide to implementation.* Thousand Oaks, CA: Corwin Press.

Milner, A. R., and T. Sondergeld. 2015. Gifted urban middle school students: The inquiry continuum and the nature of science. *National Journal of Urban Education and Practice* 8 (3): 442–461.

Morrison, V., and A. R. Milner. 2014. Literacy in support of science: A closer look at cross-curricular instructional practice. *Michigan Reading Journal* 46 (2): 42–56.

National Association for the Education of Young Children (NAEYC). 2016. Developmentally appropriate practice position statements. *www.naeyc.org/positionstatements/dap.*

Peters, E. E. 2010. Shifting to a student-centered science classroom: An exploration of teacher and student changes in perceptions and practices. *Journal of Science Teacher Education* 21 (3): 329–349.

Peters, E. E. 2012. Developing content knowledge in students through explicit teaching of the nature of science: Influences of goal setting and self-monitoring. *Science and Education* 21 (6): 881–898.

Peters, E. E., and A. Kitsantas. 2010. The effect of nature of science metacognitive prompts on science students' content and nature of science knowledge, metacognition, and self-regulatory efficacy. *School Science and Mathematics* 110: 382–396.

Popham, W. J. 2013. *Classroom assessment: What teachers need to know.* 7th ed. Upper Saddle River, NJ: Pearson.

Ritchhart, R., M. Church, and K. Morrison. 2011. *Making thinking visible: How to promote engagement, understanding, and independence for all learners.* San Francisco, CA: Jossey-Bass.

Sondergeld, T. A., C. A. Bell, and D. M. Leusner. 2010. Understanding how teachers engage in formative assessment. *Teaching and Learning* 24 (2): 72–86.

Zimmerman, B. J. 2000. Attaining self-regulation: A social-cognitive perspective. In *Handbook of self-regulation,* ed. M. Boekaerts, P. Pintrich, and M. Zeidner, 13–39. San Diego: Academic Press.

PART 2

NATURAL HAZARDS

STEM ROAD MAP MODULE

NATURAL HAZARDS MODULE OVERVIEW

Andrea R. Milner, Vanessa B. Morrison, Janet B. Walton, Carla C. Johnson, and Erin Peters-Burton

THEME: Cause and Effect

LEAD DISCIPLINE: Science

MODULE SUMMARY

In this module, students learn about the effects of natural hazards on people, communities, and the environment and consider how threats to human safety from natural hazards can be minimized. They also explore the economic effects of natural hazards from the perspectives of human and natural resources. Student teams are challenged to create a plan for how people can prepare for a natural hazard to minimize its impacts on the environment and on humans (adapted from Koehler, Bloom, and Milner 2015).

ESTABLISHED GOALS AND OBJECTIVES

The goal of this module is for students to understand and demonstrate their knowledge about the influence of natural hazards on people and on other animals. At the conclusion of this module, students will be able to do the following:

- Identify various natural hazards

- Identify the basic causes of natural hazards

- Use technology to gather research information and communicate

- Identify ways that natural hazards can impact people and communities

- Identify features of structures designed to withstand earthquakes and construct models of structures that incorporate these types of features

- Identify ways that natural hazards can impact animals' homes

- Model natural hazards

- Identify the steps of the engineering design process (EDP)

- Use the EDP to complete team projects

- Identify effective collaboration practices and reflect on their teams' efforts to collaborate

- Identify models for measuring, calculating, comparing, and evaluating numbers related to the probabilities of weather occurrences

- Identify bar graphs and infographics as ways that numbers can be displayed graphically

- Create bar graphs and infographics for data sets

- Identify ways that people and communities can prepare for natural hazards to mitigate their impacts on people and property

- Communicate information about natural hazards and natural hazard preparedness to a target audience

- Identify tall tales as a type of fictional literature and create their own tall tales

- Identify the basic parts of a story

CHALLENGE OR PROBLEM FOR STUDENTS TO SOLVE: NATURAL HAZARD PREPAREDNESS CHALLENGE

Students are challenged to work in teams to develop and communicate a plan for people to prepare for one type of natural hazard, such as a flood, tornado, earthquake, volcano, wildfire, thunderstorm, or hurricane. The plan should focus on keeping people safe if a natural hazard should strike their community. As part of this plan, students produce a public service announcement about how the community can prepare for the natural hazard.

CONTENT STANDARDS ADDRESSED IN THIS STEM ROAD MAP MODULE

A full listing with descriptions of the standards this module addresses can be found in Appendix C. Listings of the particular standards addressed within lessons are provided in a table for each lesson in Chapter 4.

STEM RESEARCH NOTEBOOK

Each student should maintain a STEM Research Notebook, which will serve as a place for students to organize their work throughout this module (see p. 12 for more general

discussion on setup and use of the notebook). All written work in the module should be included in the notebook, including records of students' thoughts and ideas, fictional accounts based on the concepts in the module, and records of student progress through the EDP. The notebooks may be maintained across subject areas, giving students the opportunity to see that although their classes may be separated during the school day, the knowledge they gain is connected. The lesson plans for this module contain STEM Research Notebook Entry sections (numbered 1–31), and templates for each notebook entry are included in Appendix A (p. 119).

Emphasize to students the importance of organizing all information in a Research Notebook. Explain to them that scientists and other researchers maintain detailed Research Notebooks in their work. These notebooks, which are crucial to researchers' work because they contain critical information and track the researchers' progress, are often considered legal documents for scientists who are pursuing patents or wish to provide proof of their discovery process.

MODULE LAUNCH

Following agreed-upon rules for discussions, hold a whole-class discussion about natural hazards, asking students questions such as the following:

- What are natural hazards?
- Are there different types of natural hazards?
- What kinds of natural hazards are there?
- What causes natural hazards?
- Can people make or cause natural hazards?
- Where and when have you seen natural hazards?

This discussion gives students an opportunity to express their conceptions of natural hazards and the influence of natural hazards. Show a video about natural hazards such as "Natural Disasters" at *www.youtube.com/watch?v=_smJ13x90oM*. Then, hold a class discussion about the various natural hazards featured in this video.

PREREQUISITE SKILLS FOR THE MODULE

Students enter this module with a wide range of preexisting skills, information, and knowledge. Table 3.1 (p. 26) provides an overview of prerequisite skills and knowledge that students are expected to apply in this module, along with examples of how they apply this knowledge throughout the module. Differentiation strategies are also provided for students who may need additional support in acquiring or applying this knowledge.

Table 3.1. Prerequisite Key Knowledge and Examples of Applications and Differentiation Strategies

Prerequisite Key Knowledge	Application of Knowledge by Students	Differentiation for Students Needing Additional Support
Science • Understand cause and effect.	*Science* • Determine how natural hazards affect humans, communities, and animals' homes.	*Science* • Provide demonstrations of cause and effect (e.g., dropping egg [cause] and observing breakage [effect]), emphasizing that cause is why something happens, effect is what happens. • Read aloud picture books to class and have students identify cause-and-effect sequences. • Create a class T-chart to record causes and related effects students observe in the classroom, in nature, and in literature.
Mathematics • Demonstrate number sense.	*Mathematics* • Measure, calculate, compare, and evaluate numbers when exploring natural hazards.	*Mathematics* • Model measurement techniques using standard and nonstandard units of measurement. • Read aloud nonfiction texts about temperature, rainfall, wind, and measurement. • Provide opportunities for students to practice measurement in a variety of settings (e.g., in the classroom and outdoors).
Language and Inquiry Skills • Visualize. • Make predictions. • Record ideas and information using words and pictures. • Ask and respond to questions.	*Language and Inquiry Skills* • Make and confirm or reject predictions. • Share thought processes through keeping a notebook, asking and responding to questions, and using the engineering design process.	*Language and Inquiry Skills* • As a class, make predictions when reading fictional texts. • Model the process of using information and prior knowledge to use predictions. • Provide samples of notebook entries.
Speaking and Listening • Participate in group discussions.	*Speaking and Listening* • Engage in collaborative group discussions in the development of natural hazard plans and about how to communicate those plans.	*Speaking and Listening* • Model speaking and listening skills. • Create a class list of good listening and good speaking skills. • Read picture books that feature collaboration and teamwork.

3

POTENTIAL STEM MISCONCEPTIONS

Students enter the classroom with a wide variety of prior knowledge and ideas, so it is important to be alert to misconceptions, or inappropriate understandings of foundational knowledge. These misconceptions can be classified as one of several types: "preconceived notions," opinions based on popular beliefs or understandings; "nonscientific beliefs," knowledge students have gained about science from sources outside the scientific community; "conceptual misunderstandings," incorrect conceptual models based on incomplete understanding of concepts; "vernacular misconceptions," misunderstandings of words based on their common use versus their scientific use; and "factual misconceptions," incorrect or imprecise knowledge learned in early life that remains unchallenged (NRC 1997, p. 28). Misconceptions must be addressed and dismantled for students to reconstruct their knowledge, and therefore teachers should be prepared to take the following steps:

- *Identify students' misconceptions.*

- *Provide a forum for students to confront their misconceptions.*

- *Help students reconstruct and internalize their knowledge, based on scientific models. (NRC 1997, p. 29)*

Keeley and Harrington (2010) recommend using diagnostic tools such as probes and formative assessment to identify and confront student misconceptions and begin the process of reconstructing student knowledge. Keeley's *Uncovering Student Ideas in Science* series contains probes targeted toward uncovering student misconceptions in a variety of areas and may be a useful resource for addressing student misconceptions in this module.

Some commonly held misconceptions specific to lesson content are provided with each lesson so that you can be alert for student misunderstanding of the science concepts presented and used during this module. The American Association for the Advancement of Science has also identified misconceptions that students frequently hold regarding various science concepts (see the links at *http://assessment.aaas.org/topics*).

SRL PROCESS COMPONENTS

Table 3.2 (p. 28) illustrates some of the activities in the Natural Hazards module and how they align with the self-regulated learning (SRL) process before, during, and after learning.

Table 3.2. SRL Process Components

Learning Process Components	Examples From Natural Hazards Module	Lesson Number and Learning Component
BEFORE LEARNING		
Motivates students	Students brainstorm about natural hazards before watching a video on the subject.	Lesson 1, Introductory Activity/Engagement
Evokes prior learning	While watching a video, students document their own experiences with natural hazards.	Lesson 1, Introductory Activity/Engagement
DURING LEARNING		
Focuses on important features	Students participate in the Earthquake Shake activity, in which they simulate earthquake conditions and observe the effects of the earthquake on various structures. Students use the most earthquake-resistant designs to identify important design features.	Lesson 2, Activity/Exploration
Helps students monitor their progress	Students create simulated earthquakes and earthquake-resistant structures using the Define, Learn, Plan, Try, Test, and Decide steps of the engineering design process, and then share their products. During the Test step, students decide whether to improve their designs based on the structures' earthquake resistance.	Lesson 2, Activity/Exploration
AFTER LEARNING		
Evaluates learning	Students present public service announcements about how to prepare for a natural disaster and receive peer feedback to improve their products before video recording them for viewing by other students and parents.	Lesson 3, Explanation
Takes account of what worked and what did not work	The whole class discusses and analyzes strengths and weaknesses of each group's natural hazard preparedness plan. Groups can meet to improve and adapt their plans based on discussion.	Lesson 3, Elaboration/Application of Knowledge

STRATEGIES FOR DIFFERENTIATING INSTRUCTION WITHIN THIS MODULE

For the purposes of this curriculum module, differentiated instruction is conceptualized as a way to tailor instruction—including process, content, and product—to various student needs in your class. A number of differentiation strategies are integrated into lessons across the module. The problem- and project-based learning approach used in the lessons is designed to address students' multiple intelligences by providing a variety of entry points and methods to investigate the key concepts in the module (for example, investigating solar power from the perspectives of science and social issues via scientific inquiry, literature, journaling, and collaborative design). Differentiation strategies for students needing support in prerequisite knowledge can be found in Table 3.1 (p. 26). You are encouraged to use information gained about student prior knowledge during introductory activities and discussions to inform your instructional differentiation. Strategies incorporated into this lesson include flexible grouping, varied environmental learning contexts, assessments, compacting, tiered assignments and scaffolding, and mentoring. The following websites may be helpful resources for differentiated instruction:

- *http://steinhardt.nyu.edu/scmsAdmin/uploads/005/120/Culturally%20Responsive%20 Differentiated%20Instruction.pdf*

- *http://educationnorthwest.org/sites/default/files/12.99.pdf*

Flexible Grouping. Students work collaboratively in a variety of activities throughout this module. Grouping strategies you might employ include using student-led grouping, grouping students according to ability level or common interests, grouping students randomly, or grouping them so that students in each group have complementary strengths (for instance, one student might be strong in mathematics, another in art, and another in writing).

Varied Environmental Learning Contexts. Students have the opportunity to learn in various contexts throughout the module, including alone, in groups, in quiet reading and research-oriented activities, and in active learning in inquiry and design activities. In addition, students learn in a variety of ways, including through doing inquiry activities, journaling, reading texts, watching videos, participating in class discussion, and conducting web-based research.

Assessments. Students are assessed in a variety of ways throughout the module, including individual and collaborative formative and summative assessments. Students have the opportunity to produce work via written text, oral and media presentations, and modeling. You may choose to provide students with additional choices of media for their products (for example, PowerPoint presentations, posters, or student-created websites or blogs).

Compacting. Based on student prior knowledge, you may wish to adjust instructional activities for students who exhibit prior mastery of a learning objective. Since student work in science is largely collaborative throughout the module, this strategy may be most appropriate for mathematics, social studies, or ELA activities. You may wish to compile a classroom database of research resources and supplementary readings for different reading levels and on a variety of subjects related to the module's topic to provide opportunities for students to undertake independent reading.

Tiered Assignments and Scaffolding. Based on your awareness of student ability, understanding of concepts, and mastery of skills, you may wish to provide students with variations on activities by adding complexity to assignments or providing more or fewer learning supports for activities throughout the module. For instance, some students may need additional support in identifying key search words and phrases for web-based research or may benefit from cloze sentence handouts to enhance vocabulary understanding. Other students may benefit from expanded reading selections and additional reflective writing or from working with manipulatives and other visual representations of mathematical concepts. You may also work with your school librarian to compile a classroom database of research resources and supplementary readings for different reading levels and on a variety of topics related to the module challenge to provide opportunities for students to undertake independent reading. You may find the following website on scaffolding strategies helpful: *www.edutopia.org/blog/scaffolding-lessons-six-strategies-rebecca-alber.*

Mentoring. As group design teamwork becomes increasingly complex throughout the module, you may wish to have a resource teacher, older student, or volunteer work with groups that struggle to stay on task and collaborate effectively.

STRATEGIES FOR ENGLISH LANGUAGE LEARNERS

Students who are developing proficiency in English language skills require additional supports to simultaneously learn academic content and the specialized language associated with specific content areas. WIDA (2012) has created a framework for providing support to these students and makes available rubrics and guidance on differentiating instructional materials for English language learners (ELLs). In particular, ELL students may benefit from additional sensory supports such as images, physical modeling, and graphic representations of module content, as well as interactive support through collaborative work. This module incorporates a variety of sensory supports and offers ongoing opportunities for ELL students to work collaboratively.

When differentiating instruction for ELL students, you should carefully consider the needs of these students as you introduce and use academic language in various language domains (listening, speaking, reading, and writing) throughout this module. To adequately differentiate instruction for ELL students, you should have an understanding

of the proficiency level of each student. The following five overarching preK–5 WIDA learning standards are relevant to this module:

- Standard 1: Social and Instructional Language. Focus on following directions, personal information, collaboration with peers.

- Standard 2: The Language of Language Arts. Focus on nonfiction, fiction, sequence of story, elements of story.

- Standard 3: The Language of Mathematics. Focus on basic operations, number sense, interpretation of data, patterns.

- Standard 4: The Language of Science. Focus on forces in nature, scientific process, Earth and sky, living and nonliving things, organisms and environment, weather.

- Standard 5: The Language of Social Studies. Focus on community workers, homes and habitats, jobs and careers, representations of Earth (maps and globes).

SAFETY CONSIDERATIONS FOR THE ACTIVITIES IN THIS MODULE

The safety precautions associated with each investigation are based in part on the use of the recommended materials and instructions, legal safety standards, and better professional safety practices. Selection of alternative materials or procedures for these investigations may jeopardize the level of safety and therefore is at the user's own risk. Remember that an investigation includes three parts: (1) setup, in which you prepare the materials for students to use; (2) the actual hands-on investigation, in which students use the materials and equipment; and (3) cleanup, in which you or the students clean the materials and put them away for later use. The safety procedures for each investigation apply to all three parts. For more general safety guidelines, see the Safety in STEM section in Chapter 2 (p. 18).

We also recommend that you go over the safety rules that are included as part of the safety acknowledgment form with your students before beginning the first investigation. Once you have gone over these rules with your students, have them sign the safety acknowledgment form. You should also send the form home with students for parents or guardians to read and sign to acknowledge that they understand the safety procedures that must be followed by their children. A sample elementary safety acknowledgment form can be found on the National Science Teaching Association's Safety Portal at *http://static.nsta.org/pdfs/SafetyAcknowledgmentForm-ElementarySchool.pdf*.

DESIRED OUTCOMES AND MONITORING SUCCESS

The desired outcome for this module is outlined in Table 3.3, along with suggested ways to gather evidence to monitor student success. For more specific details on desired outcomes, see the Established Goals and Objectives sections for the module (p. 23) and individual lessons.

Table 3.3. Desired Outcome and Evidence of Success in Achieving Identified Outcome

Desired Outcome	Evidence of Success	
	Performance Tasks	Other Measures
Students will understand and can demonstrate their knowledge about the influence of natural hazards on people and on animals' homes.	• Students complete a variety of investigations related to natural hazards. • Student teams develop and communicate natural hazard preparedness plans. • Students each maintain a STEM Research Notebook that includes what they have learned, responses to questions, and observations.	Students are assessed using the Observation, STEM Research Notebook, and Participation Rubric.

ASSESSMENT PLAN OVERVIEW AND MAP

Table 3.4 provides an overview of the major group and individual *products* and *deliverables*, or things that students will produce in this module, that constitute the assessment for this module. See Table 3.5 (p. 34) for a full assessment map of formative and summative assessments in this module.

Table 3.4. Major Products and Deliverables in Lead Discipline for Groups and Individuals

Lesson	Major Group Products and Deliverables	Major Individual Products and Deliverables
1	• Vortex Bottles	• STEM Research Notebook entries #1–11 • Weather Tall Tale • Lesson assessment
2	• Earthquake Shake structures • Group presentations of Earthquake Shake structures • Hazard Sleuths research and poster	• STEM Research Notebook entries #12–22 • "Animals in a Natural Hazard" story (creative writing) • Lesson assessment
3	• Community infographics • Our Natural Hazard Preparedness Plans public service announcements	• STEM Research Notebook entries #23–31 • Lesson assessment

Table 3.5. Assessment Map for Natural Hazards Module

Lesson	Assessment	Group/ Individual	Formative/ Summative	Lesson Objective Assessed
1	STEM Research Notebook *entries*	Individual/ Group	Formative	• Identify various natural hazards. • Identify the causes of various natural hazards. • Identify ways that mathematics can be used to describe natural phenomena. • Use a model to explain the behavior of debris in a tornado. • Identify several impacts natural hazards can have on people and communities.
1	Vortex Bottle Investigation *performance task*	Group	Formative	• Identify examples of physical models. • Create a model of tornado winds. • Use a model to explain the behavior of debris in a tornado.
1	Weather Tall Tale *creative writing rubric*	Individual/ Group	Formative	• Identify the characteristics of a tall tale. • Identify the basic parts of a story. • Create a tall tale related to weather events.
1	Lesson assessment	Individual	Formative	• Identify various natural hazards. • Identify the causes of various natural hazards. • Identify examples of physical models. • Understand that mathematical models are used to predict weather. • Identify several impacts natural hazards can have on people and communities.

Continued

Table 3.5. (*continued*)

Lesson	Assessment	Group/ Individual	Formative/ Summative	Lesson Objective Assessed
2	STEM Research Notebook *entries*	Individual/ Group	Formative	• Explain that the movement of tectonic plates can cause natural hazards. • Identify natural hazards associated with movements of tectonic plates. • Identify several impacts that natural hazards associated with the movement of tectonic plates can have on people and communities. • Identify the influence natural hazards can have on animals, with an emphasis on animals' homes. • Use bar graphs to model earthquake data and identify geographic patterns.
2	Earthquake Shake *structures and group presentations*	Group	Formative	• Identify the steps of the engineering design process (EDP). • Use the EDP to complete a group task. • Understand that design features of structures can help protect people during natural hazard events and apply that understanding to create structures designed to withstand a simulated earthquake.
2	Hazard Sleuths *research and poster*	Group	Formative	• Use technology to gather research information and communicate about natural hazards. • Identify several impacts that natural hazards associated with the movement of tectonic plates can have on people and communities. • Identify several ways that people can remain safe during a natural hazard occurrence. • Communicate information about natural hazards in a visual format.
2	"Animals in a Natural Hazard" story *creative writing rubric*	Individual	Formative	• Identify the influence natural hazards can have on animals, with an emphasis on animals' homes. • Identify several ways that people can remain safe during a natural hazard occurrence.

Continued

Table 3.5. (*continued*)

Lesson	Assessment	Group/ Individual	Formative/ Summative	Lesson Objective Assessed
2	Lesson assessment	Individual	Formative	• Identify several impacts that natural hazards associated with the movement of tectonic plates can have on people and communities.
3	STEM Research Notebook *prompts*	Individual/ Group	Formative	• Identify impacts of natural hazards on people and the environment. • Create a preparedness plan that can mitigate the impacts of a natural hazard on people and the environment. • Use technology tools to gather data about natural hazards.
3	Community infographics *performance task*	Group	Formative	• Understand that community characteristics can be expressed numerically and in text. • Organize numerical and textual information about students' communities in an infographic.
3	Our Natural Hazard Preparedness Plans public service announcements *performance task*	Group	Summative	• Identify impacts of natural hazards on people and the environment. • Create a preparedness plan that can mitigate the impacts of a natural hazard on people and the environment. • Communicate understanding of natural hazard preparedness through a PSA. • Understand that community characteristics can be expressed numerically and in text. • Use technology to communicate about natural hazards.
3	Lesson assessment	Individual	Summative	• Identify impacts of natural hazards on people and the environment.

MODULE TIMELINE

Tables 3.6–3.10 (pp. 37–40) provide lesson timelines for each week of the module. These timelines are provided for general guidance only and are based on class times of approximately 30 minutes.

Table 3.6. STEM Road Map Module Schedule for Week One

Day 1	Day 2	Day 3	Day 4	Day 5
Lesson 1 *Let's Explore Natural Hazards*	*Lesson 1* *Let's Explore Natural Hazards*	*Lesson 1* *Let's Explore Natural Hazards*	*Lesson 1* *Let's Explore Natural Hazards*	*Lesson 1* *Let's Explore Natural Hazards*
• Launch the module by holding a group discussion about natural hazards and showing a video. • Introduce the module challenge. • Introduce a current natural hazard.	• The class classifies natural hazards according to their causes. • Conduct an interactive read-aloud of *Violent Weather: Thunderstorms, Tornadoes, and Hurricanes*, by Andrew Collins. • Begin vocabulary chart.	• Show a video about tornadoes. • Discuss wind patterns in tornadoes. • Introduce the Predict, Observe, Explain (POE) process. • Introduce modeling. • Introduce the use of numbers to describe natural hazards, and conduct an interactive read-aloud of pages 4–21 of *Hurricanes (Real World Math: Natural Disasters series)*, by Barbara A. Somervill.	• Conduct Vortex Bottle Investigation (Predict and Observe). • Discuss weather forecasting and probabilities. • Conduct an interactive read-aloud of *That's a Possibility!: A Book About What Might Happen*, by Bruce Goldstone.	• Conclude Vortex Bottle Investigation (Explain). • Introduce tall tales. • Conduct interactive read-aloud of *Cloudy With a Chance of Meatballs*, by Judi Barrett. • Students begin planning and writing their Weather Tall Tales

Table 3.7. STEM Road Map Module Schedule for Week Two

Day 6	Day 7	Day 8	Day 9	Day 10
Lesson 1 *Let's Explore Natural Hazards* • Discuss floods and conduct an interactive read-aloud of *Flood Warning* (*Let's-Read-and-Find-Out Science 2*), by Katharine Kenah. • Continue writing Weather Tall Tales.	*Lesson 1* *Let's Explore Natural Hazards* • Conduct lesson assessment. • Complete Weather Tall Tales.	*Lesson 2* *Natural Hazard Quest!* • Discuss movement of tectonic plates as a cause of earthquakes and associated natural hazards. • Conduct an interactive read-aloud of the book *Earthquakes*, by Ellen Prager. • Discuss animal habitats and the impacts of natural hazards on animals' homes. • Conduct an interactive read-aloud or show the video of *A House Is a House for Me*, by Mary Ann Hoberman	*Lesson 2* *Natural Hazard Quest!* • Discuss impacts of natural hazards on people and communities, and show and discuss before-and-after images of natural disasters. • Investigate and document financial costs associated with a natural hazard that occurred recently.	*Lesson 2* *Natural Hazard Quest!* • Introduce the engineering design process. • Create class collaboration rules and contracts. • Begin Earthquake Shake activity (Define and Learn).

Table 3.8. STEM Road Map Module Schedule for Week Three

Day 11	Day 12	Day 13	Day 14	Day 15
Lesson 2 *Natural Hazard Quest!* • Continue Earthquake Shake activity (Plan, Try, and Test). • Introduce bar graphs. • Conduct interactive read-aloud of *Lemonade for Sale*, by Stuart J. Murphy.	*Lesson 2* *Natural Hazard Quest!* • Continue Earthquake Shake activity (Plan, Try, and Test). • Students create bar graphs for current earthquake magnitudes by continent.	*Lesson 2* *Natural Hazard Quest!* • Share Earthquake Shake activity designs, and test best class designs. • Students share bar graphs of earthquake magnitudes by continent. • Begin planning and writing stories about animal homes in natural hazards.	*Lesson 2* *Natural Hazard Quest!* • Conduct research for Hazard Sleuths activity. • Introduce U.S. regions and earthquake statistics for these regions. • Continue planning and writing stories about animal homes in natural hazards.	*Lesson 2* *Natural Hazard Quest!* • Continue research for Hazard Sleuths activity. • Students create bar graphs for earthquake magnitude by U.S. region. • Continue writing stories about animal homes in natural hazards.

Table 3.9. STEM Road Map Module Schedule for Week Four

Day 16	Day 17	Day 18	Day 19	Day 20
Lesson 2 *Natural Hazard Quest!* • Create posters for Hazard Sleuths activity. • Continue writing stories about animal homes in natural hazards.	*Lesson 2* *Natural Hazard Quest!* • Create posters for Hazard Sleuths activity. • Complete stories about animal homes in natural hazards. • Conduct lesson assessment.	*Lesson 3* *Our Natural Hazard Preparedness Plans* • Introduce natural hazard preparedness through class discussion. • Introduce PSAs through class discussion and video. • Introduce numerical information about the community and infographics.	*Lesson 3* *Our Natural Hazard Preparedness Plans* • Discuss thunderstorm preparedness through an interactive read-aloud of *Flash, Crash, Rumble, and Roll* by Franklyn M. Branley. • Students create community infographics.	*Lesson 3* *Our Natural Hazard Preparedness Plans* • Introduce use of the engineering design process and storyboards for PSAs. • Class decides on types of information that should be included in PSAs (Define).

Table 3.10. STEM Road Map Module Schedule for Week Five

Day 21	Day 22	Day 23	Day 24	Day 25
Lesson 3 *Our Natural Hazard Preparedness Plans* • Conduct research for PSAs (Learn).	*Lesson 3* *Our Natural Hazard Preparedness Plans* • Create storyboards for PSAs (Plan).	*Lesson 3* *Our Natural Hazard Preparedness Plans* • Teams practice PSAs (Try). • Teams give feedback to and receive feedback from one other team (Test). • Teams decide how to improve their PSAs (Decide).	*Lesson 3* *Our Natural Hazard Preparedness Plans* • Teams present their PSAs.	*Lesson 3* *Our Natural Hazard Preparedness Plans* • Teams discuss possible improvements to PSAs based on class discussion. • Conduct lesson assessment.

RESOURCES

The media specialist can help teachers locate resources for students to view and read about natural hazards, habitats, animal homes, and related content. Special educators and reading specialists can help find supplemental sources for students needing extra support in reading and writing. Additional resources may be found online. Community resources for this module may include meteorologists, climate scientists, emergency services personnel, and public safety officials.

REFERENCES

Koehler, C., M. A. Bloom, and A. R. Milner. 2015. The STEM Road Map for grades K–2. In *STEM Road Map: A framework for integrated STEM education*, ed. C. C. Johnson, E. E. Peters-Burton, and T. J. Moore, 41–67. New York: Routledge. *www.routledge.com/products/9781138804234*.

Keeley, P., and R. Harrington. 2010. *Uncovering student ideas in physical science, volume 1: 45 new force and motion assessment probes.* Arlington, VA: NSTA Press.

National Research Council (NRC). 1997. *Science teaching reconsidered: A handbook.* Washington, DC: National Academies Press.

WIDA. 2012. 2012 amplification of the English language development standards: Kindergarten–grade 12. *https://wida.wisc.edu/teach/standards/eld*.

NATURAL HAZARDS LESSON PLANS

Andrea R. Milner, Vanessa B. Morrison, Janet B. Walton, Carla C. Johnson, and Erin Peters-Burton

Lesson Plan 1: Let's Explore Natural Hazards

In this lesson, students explore natural hazards and their causes, with the aim of understanding that various types of natural hazards occur around the world, and these hazards can be classified as those with weather-related causes and those caused by movements within Earth. Students explore tornadoes as an example of a natural hazard caused by weather.

ESSENTIAL QUESTIONS

- What are natural hazards?
- What are the different types of natural hazards?
- What causes natural hazards?
- Can people make or cause natural hazards?
- What impact do natural hazards have on people?
- What natural hazards are caused by weather?

ESTABLISHED GOALS AND OBJECTIVES

At the conclusion of this lesson, students will be able to do the following:

- Identify various natural hazards
- Identify the basic causes of natural hazards
- Identify examples of physical models
- Understand that mathematical models are used to predict weather
- Create a model of tornado winds

- Use a model to explain the behavior of debris in a tornado
- Identify several impacts natural hazards can have on people and communities
- Identify ways that mathematics can be used to describe natural phenomena
- Identify the characteristics of a tall tale
- Identify the basic parts of a story
- Create a tall tale related to weather events

TIME REQUIRED

- 7 days (approximately 30 minutes each day; see Tables 3.6–3.7, pp. 37 and 38)

MATERIALS

Required Materials for Lesson 1

- STEM Research Notebooks (1 per student; see p. 24 for STEM Research Notebook information)
- Computer with internet access for viewing videos
- 2 sheets of plain white paper (per student)
- 3 sheets of lined writing paper (per student)
- Books:
  - *Violent Weather: Thunderstorms, Tornadoes, and Hurricanes,* by Andrew Collins (National Geographic Children's Books, 2006)
  - *Hurricanes (Real World Math: Natural Disasters series),* by Barbara A. Somervill (Cherry Lake, 2012)
  - *That's a Possibility!: A Book About What Might Happen,* by Bruce Goldstone (Henry Holt and Co., 2013)
  - *Cloudy With a Chance of Meatballs,* by Judi Barrett (Atheneum Books for Young Readers, 1982)
  - *Flood Warning (Let's-Read-and-Find-Out Science 2),* by Katharine Kenah (HarperCollins, 2016)
- Chart paper
- Markers

- Map or globe

- Crayons for use in STEM Research Notebook entries (1 set per student)

- Safety glasses or indirectly vented chemical splash goggles, nonlatex aprons, and nonlatex vinyl gloves (per student)

Additional Materials for Vortex Bottle Investigation (per pair of students)

- Water (about 2 ½ cups)

- Clear 2 liter plastic bottle with label removed and a cap

- Glitter (about 2 tablespoons)

- Dishwashing liquid (about 1 teaspoon)

- 3 paper towels

SAFETY NOTES

1. All students must wear sanitized indirectly vented chemical-splash goggles, nonlatex aprons, and vinyl gloves during all phases of this inquiry activity.

2. Keep away from electrical receptacles when using water to avoid shock hazard.

3. Keep glitter away from eyes and mouths.

4. Immediately clean up any spills on the floor to avoid a slip-and-fall hazard.

5. Wash hands with soap and water after completing the cleanup phase of the activity.

CONTENT STANDARDS AND KEY VOCABULARY

Table 4.1 lists the content standards from the *Next Generation Science Standards (NGSS)*, *Common Core State Standards (CCSS)*, National Association for the Education of Young Children (NAEYC), and the Framework for 21st Century Learning that this lesson addresses, and Table 4.2 (p. 50) presents the key vocabulary. Vocabulary terms are provided for both teacher and student use. Teachers may choose to introduce some or all of the terms to students.

Table 4.1. Content Standards Addressed in STEM Road Map Module Lesson 1

NEXT GENERATION SCIENCE STANDARDS

PERFORMANCE EXPECTATIONS

- 2-ESS1-1. Use information from several sources to provide evidence that Earth events can occur quickly or slowly.

SCIENCE AND ENGINEERING PRACTICES

Analyzing and Interpreting Data

Analyzing data in K–2 builds on prior experiences and progresses to collecting, recording, and sharing observations.
- Use observations (firsthand or from media) to describe patterns in the natural world in order to answer scientific questions.

Developing and Using Models

Modeling in K–2 builds on prior experiences and progresses to include using and developing models (i.e., diagram, drawing, physical replica, diorama, dramatization, storyboard) that represent concrete events or design solutions.
- Use a model to represent relationships in the natural world.

Planning and Carrying Out Investigations

Planning and carrying out investigations to answer questions or test solutions to problems in K–2 builds on prior experiences and progresses to simple investigations, based on fair tests, which provide data to support explanations or design solutions.
- Make observations (firsthand or from media) to collect data that can be used to make comparisons.

DISCIPLINARY CORE IDEAS

LS1.C. Organization for Matter and Energy Flow in Organisms
- All animals need food in order to live and grow. They obtain their food from plants or from other animals. Plants need water and light to live and grow.

Continued

Table 4.1. (*continued*)

ESS3.A. Natural Resources

- Living things need water, air, and resources from the land, and they live in places that have the things they need. Humans use natural resources for everything they do.

ESS2.D. Weather and Climate

- Weather is the combination of sunlight, wind, snow or rain, and temperature in a particular region at a particular time. People measure these conditions to describe and record the weather and to notice patterns over time.

CROSSCUTTING CONCEPTS

Patterns

- Patterns in the natural and human designed world can be observed, used to describe phenomena, and used as evidence.

Systems and System Models

- Systems in the natural and designed world have parts that work together.

Cause and Effect

- Events have causes that generate observable patterns.

COMMON CORE STATE STANDARDS FOR MATHEMATICS

MATHEMATICAL PRACTICES

- MP1. Make sense of problems and persevere in solving them.
- MP2. Reason abstractly and quantitatively.
- MP3. Construct viable arguments and critique the reasoning of others.
- MP4. Model with mathematics.
- MP5. Use appropriate tools strategically.
- MP6. Attend to precision.
- MP7. Look for and make use of structure.
- MP8. Look for and express regularity in repeated reasoning.

MATHEMATICAL CONTENT

- 2.NBT.A.1. Understand that the three digits of a three-digit number represent amounts of hundreds, tens, and ones; e.g., 706 equals 7 hundreds, 0 tens, and 6 ones.
- 2.NBT.A.2. Count within 1,000; skip-count by 5s, 10s, and 100s.
- 2.NBT.A.3. Read and write numbers to 1,000 using base-ten numerals, number names, and expanded form.

Continued

Table 4.1. (*continued*)

COMMON CORE STATE STANDARDS FOR ENGLISH LANGUAGE ARTS

READING STANDARDS

- RI.2.1. Ask and answer such questions as *who, what, where, when, why,* and *how* to demonstrate understanding of key details in a text.

- RI.2.3. Describe the connection between a series of historical events, scientific ideas or concepts, or steps in technical procedures in a text.

- RI.2.7. Explain how specific images (e.g., a diagram showing how a machine works) contribute to and clarify a text.

- RI.2.8. Describe how reasons support specific points the author makes in a text.

- RI.2.9. Compare and contrast the most important points presented by two texts on the same topic.

WRITING STANDARDS

- W.2.1. Write opinion pieces in which they introduce the topic or book they are writing about, state an opinion, supply reasons that support the opinion, use linking words (e.g., *because, and, also*) to connect opinion and reasons, and provide a concluding statement or section.

- W.2.2. Write informative/explanatory texts in which they introduce a topic, use facts and definitions to develop points, and provide a concluding statement or section.

- W.2.6. With guidance and support from adults, use a variety of digital tools to produce and publish writing, including in collaboration with peers.

- W.2.7. Participate in shared research and writing projects (e.g., read a number of books on a single topic to produce a report; record science observations).

- W.2.8. Recall information from experiences or gather information from provided sources to answer a question.

SPEAKING AND LISTENING STANDARDS

- SL.2.1. Participate in collaborative conversations with diverse partners about *grade 2 topics and texts* with peers and adults in small and larger groups.

- SL.2.2. Recount or describe key ideas or details from a text read aloud or information presented orally or through other media.

- SL.2.3. Ask and answer questions about what a speaker says in order to clarify comprehension, gather additional information, or deepen understanding of a topic or issue.

Continued

Table 4.1. (*continued*)

NATIONAL ASSOCIATION FOR THE EDUCATION OF YOUNG CHILDREN STANDARDS

- 2.G.02. Children are provided varied opportunities and materials to learn key content and principles of science.

- 2.G.03. Children are provided varied opportunities and materials that encourage them to use the five senses to observe, explore, and experiment with scientific phenomena.

- 2.G.04. Children are provided varied opportunities to use simple tools to observe objects and scientific phenomena.

- 2.G.05. Children are provided varied opportunities and materials to collect data and to represent and document their findings (e.g., through drawing or graphing).

- 2.G.06. Children are provided varied opportunities and materials that encourage them to think, question, and reason about observed and inferred phenomena.

- 2.G.07. Children are provided varied opportunities and materials that encourage them to discuss scientific concepts in everyday conversation.

- 2.G.08. Children are provided varied opportunities and materials that help them learn and use scientific terminology and vocabulary associated with the content areas.

- 2.H.02. All children have opportunities to access technology (e.g., tape recorders, microscopes, computers) that they can use.

- 2.H.03. Technology is used to extend learning within the classroom and integrate and enrich the curriculum.

FRAMEWORK FOR 21ST CENTURY LEARNING

- Interdisciplinary Themes; Learning and Innovation Skills; Information, Media, and Technology Skills; Life and Career Skills.

Table 4.2. Key Vocabulary for Lesson 1

Key Vocabulary	Definition
data	information collected by making observations, taking measurements, or asking questions
drought	a long period of dry weather with very little rainfall
earthquake	a sudden and dangerous shaking of the ground caused by movements in Earth's crust
environment	the conditions of the natural world, including living things, that make up our surroundings
equation	a mathematical expression that says two things are equal
evacuate	to move away from a dangerous area
evidence	factual information that can be used to support a belief or theory
flood	an overflow of a large amount of water over land that is normally dry
gravity	a force that pulls objects toward Earth
hurricane	a violent storm with very strong winds that forms over tropical waters; also called *cyclone* or *typhoon*, depending on where it occurs
landslide	the sliding of large amounts of soil and rocks down a mountain or cliff resulting from a storm
mathematical model	a way to describe something using numbers
meteorologist	a scientist who studies the atmosphere and the weather
model	a representation of something that is difficult to see or understand in everyday life
natural disaster	a natural hazard that results in harm to people or property
natural hazard	an event caused by nature that has the potential to harm people and the environment
percentage	a fraction that tells parts out of 100; uses the symbol %
physical model	a representation of an object using materials to create a larger or smaller version of the object
precipitation	water particles that fall from the sky to the ground in liquid or solid form, such as rain, snow, ice, and hail
prepare	to plan and get ready ahead of time

Continued

Table 4.2. (*continued*)

Key Vocabulary	Definition
prevent	to stop something from taking place
probability	the chance or likelihood that something will happen; can be measured with numbers (e.g., 10% chance) or expressed with words (e.g., possible, likely, unlikely, impossible, certain)
thundercloud	a cloud filled with electricity that produces thunder and lightning
thunderstorm	a storm with lightning and thunder, as well as heavy rain
tornado	violently spinning winds that resemble a funnel-shaped cloud
volcano	an opening in a mountain from which hot rocks, lava, and gas are pushed out
vortex	rotating liquid or air that tends to form a cone shape
weather	the daily conditions over a particular area that include temperature, precipitation, cloud cover, and air pressure
wildfire	a destructive fire that quickly spreads through woods, forests, and brush

TEACHER BACKGROUND INFORMATION

Second graders are able to make connections across multiple content areas (STEM and ELA), as well as the various developmental domains (physical, social and emotional, personality, cognitive, and language). Incorporating students' prior knowledge with developmentally appropriate instruction will enable them to make these connections. Throughout this module, you should support and facilitate the advancement of these content areas and developmental domains within each student. For information about how formative assessments can be used to connect students' prior experiences with classroom instruction, see the STEM Teaching Tools resource "Making Science Instruction Compelling for All Students: Using Cultural Formative Assessment to Build on Learner Interest and Experience" at *http://stemteachingtools.org/pd/sessionc.*

Natural Hazards

This module focuses on natural hazards and their effects on people and the environment. The terms natural hazard and natural disaster have different meanings to Earth scientists. Natural hazards are considered natural phenomena, such as tornadoes and hurricanes, that *have the potential* to affect people and property, while natural disasters are natural hazards that actually *do* affect people and property (see the U.S. Geological Survey's "EarthWord—Hazard" page at *www.usgs.gov/news/earthword-hazard* for more information). This is a fine distinction that you need not introduce to students, but both terms are used in this module. For the purposes of this module, natural hazard is generally used to refer to the natural phenomena, while natural disaster is used to refer to the human experience of the event, although these terms can be used interchangeably for the purposes of classroom discussion.

Natural hazards are of two primary types: those associated with movements within Earth and those associated with weather events. Natural hazards caused by movements within Earth include earthquakes, volcanic eruptions, and tsunamis. Hazards of this type are difficult to predict, although new technologies are evolving to detect seismic activity. Natural hazards caused by weather events include tornadoes, hurricanes, blizzards, extremely heavy rains, drought, extreme heat, and extreme cold weather. Several types of natural disasters are supplementary to natural hazards, such as floods and landslides caused by heavy rains. For more information about various kinds of natural hazards, see the following websites:

- *www.earthtimes.org/encyclopaedia/environmental-issues/natural-disasters*

- *www.usgs.gov/science/science-explorer/Natural+Hazards*

- *https://earthquake.usgs.gov*

- *www.usgs.gov/natural-hazards/landslide-hazards*

- *https://volcanoes.usgs.gov/index.html*

- *www.nasa.gov/content/esd-natural-disasters*

Students discuss and explore a variety of natural hazards in this module; however, you may wish to place particular emphasis on one kind of hazard that is common in your geographic area and have students create their preparedness plans for this hazard. The American Red Cross provides an interactive map that identifies the most common natural disasters for each region of the United States at *www.redcross.org/get-help/how-to-prepare-for-emergencies/common-natural-disasters-across-us.html*.

This lesson focuses on natural hazards with weather-related causes, using tornadoes as an example. Most tornadoes originate from rotating thunderstorms and are rapidly rotating columns of air that reach from the thunderstorms to the ground. Wind speeds of the most extreme tornadoes have exceeded 300 miles per hour (mph). They can cause devastating damage to structures, uproot trees, and move vehicles and other property. Conditions that favor the formation of tornadoes include thunderstorms, conditions that lift moist air such as cold fronts, an unstable atmosphere (the temperature decreases rapidly with increasing height), and areas with strong winds that are turning in a clockwise direction.

Tornadoes are traditionally envisioned as funnel-shaped clouds that form a vortex, or spinning winds that are empty in the center. Tornadoes can be formed from one vortex or can contain multiple vortices. In this lesson, students will create a simple model of a vortex. This model is not intended to accurately model a tornado, but rather to allow students to explore the motion associated with tornado winds and observe that matter can move down through the center of a tornado.

Tornadoes are difficult to predict in advance since they form from a confluence of atmospheric factors; however, computer modeling can be used to identify where weather conditions might be favorable to tornado formation. The south-central portion of the United States, called "Tornado Alley," and Florida are areas with relatively high tornado frequencies. In Florida, tornadoes most frequently form in the late fall (October through December), while Tornado Alley experiences the most tornadoes in the late spring and early fall. Meteorologists at the National Oceanic and Atmospheric Administration's (NOAA's) Storm Prediction Center monitor the atmosphere for severe thunderstorms and conditions favorable to tornado formation. They may issue tornado *watches* based on these conditions. Tornado *warnings* are issued by local National Weather Service offices when a tornado has been sighted in an area or identified by weather radar.

Tornadoes' strengths are assigned ratings from 0 to 5 using the Enhanced Fujita Scale or EF Scale (introduced as a revision of the original Fujita Scale in 2007). This scale uses wind speed estimates and also takes into account observed damage based on variables such as types of structures. An EF0 tornado has wind speeds between 65 and 85 mph and only light damage, while an EF5 tornado has wind speeds over 200 mph with devastating damage. More information about tornadoes is available on the following websites:

- *www.ready.gov/tornadoes*

- *www.kids.nationalgeographic.com/explore/science/tornado/#tornado.jpg*

- *www.weatherwizkids.com/weather-tornado.htm*

- *www.nssl.noaa.gov/education/svrwx101/tornadoes/forecasting*

Optional Classroom Technology Tools

You might consider introducing mobile apps that allow tracking of natural hazard events and provide emergency preparedness information, such as the following:

- Apps available from the U.S. Red Cross, such as Monster Guard (geared toward emergency preparedness for children ages 7–11) and others for earthquakes, tornadoes, hurricanes, and floods

- Apps available for Apple mobile devices, such as Seismograph, Tremor Tracker, Dynamic Plates, and Volcanoes: Map, Alerts & Ash (note that these or similar apps are also be available for Android devices)

In addition, the LEGO Education WeDo platform provides a variety of activities related to natural hazards that you may wish to incorporate into the module if your school has these kits available. These include the following:

- *https://education.lego.com/en-us/lessons/wedo-2-science/robust-structures*
- *https://education.lego.com/en-us/lessons/wedo-2-science/prevent-flooding*
- *https://education.lego.com/en-us/lessons/wedo-2-computational-thinking/volcano-alert*
- *https://education.lego.com/en-us/lessons/wedo-2-science/drop-and-rescue*
- *https://education.lego.com/en-us/lessons/wedo-2-science/hazard-alarm*

Career Connections

You may wish to introduce careers associated with weather and natural disaster preparedness, such as the following (see Koehler, Bloom, and Milner 2015):

- Climatologist
- Ecologist
- Environmental engineer
- Geographer
- Materials engineer
- Meteorologist
- Urban planner

For more information about these and other careers, see the Bureau of Labor Statistics' *Occupational Outlook Handbook* at *www.bls.gov/ooh/home.htm*.

In this module, students are introduced to the idea that engineers work together in teams with those in other STEM careers to solve problems. Students experience working in teams and in pairs as they progress through a simple scientific process including predicting, observing, and explaining phenomena related to natural hazards in this lesson. This introduction to teamwork sets the stage for students' use of the engineering design process (EDP) later in the module.

Know, Learning, Evidence, Wonder, Scientific Principle (KLEWS) Charts

You will track student knowledge on Know, Learning, Evidence, Wonder, Scientific Principle (KLEWS) charts throughout this module. These charts are used to access and assess student prior knowledge, encourage students to think critically about the topic under discussion, and track student learning throughout the module. Using KLEWS charts challenges students to connect evidence and scientific principles with their learning. Be sure to list the topic at the top of each chart. The charts should consist of five columns, one for each KLEWS component. It may be helpful to post these charts in a prominent place in the classroom so that students can refer to them throughout the module. Students will write their personal ideas and reflections in their STEM Research Notebooks. For more information about KLEWS charts, see the January 2006 National Science Teaching Association *Web-News Digest* article "Evidence Helps the KWL get a KLEW" at *www.nsta.org/publications/news/story.aspx?id=51519* or the February 2015 *Science and Children* article "Methods and Strategies: KLEWS to Explanation-Building in Science" at *www.nsta.org/store/product_detail.aspx?id=10.2505/4/sc15_052_06_66*.

Interactive Read-Alouds

This module also uses interactive read-alouds to engage students, access their prior knowledge, develop their background knowledge, and introduce topical vocabulary. These read-alouds expose children to teacher-read literature that may be beyond their independent reading levels but is consistent with their listening level. Interactive read-alouds may incorporate a variety of techniques, and you can find helpful information regarding these techniques at the following websites:

- *www.readingrockets.org/article/repeated-interactive-read-alouds-preschool-and-kindergarten*

- *www.k5chalkbox.com/interactive-read-aloud.html*

- *www.readwritethink.org/professional-development/strategy-guides/teacher-read-aloud-that-30799.html*

In general, interactive read-alouds provide opportunities for students to share prior knowledge and experiences, interact with the text and concepts introduced therein, launch conversations about the topics introduced, construct meaning, make predictions, and draw comparisons. You may wish to mark places within the texts to pause and ask students to share their experiences, predictions, or other ideas. Each reading experience should focus on an ongoing interaction between students and the text, making time to do the following:

- Allow students to share personal stories throughout the reading

- Ask students to predict throughout the story
- Allow students to add new ideas from the book to the KLEWS chart and their STEM Research Notebooks
- Allow students to add new words from the book to the vocabulary chart and their STEM Research Notebooks

The materials list for each lesson includes the books for interactive read-alouds that you will use in that lesson. A list of suggested books for additional reading can be found at the end of this chapter (see p. 112).

Working With Large Numbers

In this module's mathematics and social studies connections, students consider numerical data associated with natural hazards. Some of these data, such as homes without power, people evacuated, and financial costs of rebuilding and repairs, may be expressed in large numbers, including hundreds of thousands and millions. You should therefore be prepared to discuss place values and large numbers with students, and you may wish to prepare a chart of place values to display in the classroom. The following websites provide ideas and additional resources for introducing large numbers to early elementary students:

- *www.mathcoachscorner.com/2012/08/place-value-reading-large-numbers*
- *www.mathgeekmama.com/books-teach-place-value-large-numbers*

COMMON MISCONCEPTIONS

Students will have various types of prior knowledge about the concepts introduced in this lesson. Table 4.3 outlines some common misconceptions students may have concerning these concepts. Because of the breadth of students' experiences, it is not possible to anticipate every misconception that students may bring as they approach this lesson. Incorrect or inaccurate prior understanding of concepts can influence student learning in the future, however, so it is important to be alert to misconceptions such as those presented in the table.

Table 4.3. Common Misconceptions About the Concepts in Lesson 1

Topic	Student Misconception	Explanation
Natural hazards	Tornadoes only occur in the Midwest.	Although tornadoes are most likely to occur in Midwestern states, they can occur anywhere.
	A small amount of water from flooding is not dangerous.	As little as 6 inches of moving water can knock people off their feet, and about 2 feet of water will cause a car to float.
	There is nothing we can do to protect ourselves from natural hazards.	While we cannot prevent hazards such as tornadoes, hurricanes, earthquakes, and volcanic eruptions, we can take measures to protect human life and property in areas likely to be affected by those hazards. These include designing homes and buildings to withstand hurricanes' high winds and earthquake tremors, providing basements or safe rooms where people can go during tornado warnings, and evacuating people from areas in the path of a hurricane.
Weather	We can predict weather in the long term by looking at animals' behaviors and their coats.	We cannot predict weather by animals' behaviors and the thickness of their fur. Weather can be predicted only by observing factors in the atmosphere, so it is difficult to make long-range weather predictions.
	Wind is caused by cold weather.	Wind is caused by uneven heating of Earth's surface and pockets of air with different amounts of heat.
	Clouds block or slow down wind.	Clouds are large collections of tiny water droplets or ice crystals that are light enough to float in the air; they do not block the wind but can be moved by the wind.
	Snow and ice create cold temperatures.	Snow and ice are caused by cold temperatures, which freeze moisture in the atmosphere.

PREPARATION FOR LESSON 1

Review the Teacher Background Information section (p. 51), assemble the materials for the lesson, duplicate the student handouts, and preview the videos recommended in the Learning Components section below. Present students with their STEM Research Notebooks and explain how they will be used (see p. 24). Templates for the STEM Research Notebook are provided in Appendix A, and an Observation, STEM Research Notebook, and Participation Rubric is provided in Appendix B.

You may wish to use the template for STEM Research Notebook Entry #5 throughout the module for students to record definitions and draw illustrations of key vocabulary words. The template provides space for three words. If you plan to introduce more than five vocabulary words in a lesson, you should make multiple copies of the template for each student.

Identify a natural hazard that has affected an area somewhere in the world recently. If possible, you may wish to choose a natural hazard that has affected your geographic area. The class will investigate this event and track its impacts throughout the module. You should have on hand the location of the event, some basic facts, and some pictures to share with students. You should also have available an estimate of the number of people affected by the disaster and some other numerical facts (e.g., inches of rain, days or hours in which the event occurred, homes without power, people evacuated, estimated costs to clean up, estimated lost wages). Create a written narrative about the event appropriate to your students' reading level, including numerical facts about the event within the text. Duplicate this narrative for each student. See the questions on STEM Research Notebook Entry #2 for additional information you should research beforehand and be prepared to discuss with the class.

For the mathematics connection (see p. 64), have a copy of a printed weather report or a video of a weather report that includes varying probabilities for precipitation on multiple days.

LEARNING COMPONENTS
Introductory Activity/Engagement

Connection to the Challenge: Begin each day of this lesson by directing students' attention to the module challenge, the Natural Hazard Preparedness Challenge:

> *Your town's leaders want to be sure that people will be safe in case a natural hazard should strike your town. They have asked your class to create preparedness plans for natural hazards that will keep people safe during these events. You and your team are challenged to create a plan for the community for one natural hazard and an advertisement that lets people know about your plan.*

Tell students that they will learn about different kinds of natural hazards and how they can affect people as well as animals. To do this, they need to learn about the kinds of natural hazards that occur around the world and in your area and about local animal habitats and homes. Hold a brief class discussion of how students' learning in the previous days' lessons contributed to their ability to complete the challenge. You may wish to create a class list of key ideas on chart paper.

Science Class and ELA Connection: Introduce the module by holding a class discussion about natural hazards. Put chart paper on the wall and keep it up throughout the module for students to refer to. Through whole-class discussion, have students share their conceptions of what natural hazards are and what the influences of natural hazards are. Following agreed-upon rules for discussions, ask students the following questions:

- What are natural hazards?

- Are there different types of natural hazards?

- What types of natural hazards are there?

- What causes natural hazards?

- Can people make or cause natural hazards?

- Where and when have you seen natural hazards?

As students share their ideas, chart the responses in the Know column of a KLEWS chart. Then, ask students what questions they have about the natural hazards, recording these questions in the Wonder column.

Introduce students to the STEM Research Notebooks they will use throughout the module. Explain to students that scientists and people in other STEM careers use research notebooks to track their work. Then, show a video about natural hazards, such as "Natural Disasters" at *www.youtube.com/watch?v=_smJ13x90oM*, and hold a class discussion about the various natural hazards featured in the video.

STEM Research Notebook Entry #1

Have students document their own personal experiences with natural hazards or memorable weather events in their STEM Research Notebooks, using both words and pictures.

Social Studies and Mathematics Connections: Have students name natural hazards and create a class list. Ask students if they think some natural hazards occur more often in certain places. Students should understand that natural hazards often occur where there are specific landforms (e.g., volcanoes and earthquakes) or weather patterns (e.g., tornadoes occur most often in the central United States, and hurricanes most frequently affect areas along ocean coastlines). Help students locate on the map or globe the regions

throughout the United States where these natural hazards take place. Be sure to emphasize that natural hazards can occur at other places around the world (e.g., tsunamis caused by underwater earthquakes in Japan in 2011 and in Indonesia in 2018). Explain that the same kind of storms may be called *hurricanes, cyclones,* or *typhoons,* depending on where in the world they occur.

Introduce the class to a natural hazard that was in the news recently. Help students identify the location where the natural hazard occurred on the map or globe. Have three or four students provide estimates of the number of people affected, and work as a class to find an average of students' estimates. Then, provide information about the actual number of people affected. Work as a class to find the difference between the actual number of people affected and students' estimates.

Throughout the module, you should discuss with the class the impacts this natural disaster event is having on the community in which it occurred (e.g., economic, business, health, and environmental impacts). Discuss information about the disaster (see STEM Research Notebook Entry #2).

STEM Research Notebook Entry #2

Have students record this information about a current natural disaster in their notebooks.

Activity/Exploration

Science Class and ELA Connection: Make a list on the board of various types of natural hazards (be sure to include, at a minimum, tornadoes, hurricanes, volcanoes, earthquakes, and floods). Next, ask students to identify what causes each type of disaster (i.e., weather or movement within Earth's surface) and record this on the list.

STEM Research Notebook Entry #3

Have students complete a T-chart classifying natural disasters according to their causes. The column headers of the chart are the causes (weather and movements within Earth), and students should enter each of the natural hazards listed on the board under the appropriate cause.

Next, ask students how thunderstorms, tornadoes, and hurricanes are the same and different, documenting their ideas on a KLEWS chart. Then, conduct an interactive read-aloud of *Violent Weather: Thunderstorms, Tornadoes, and Hurricanes,* by Andrew Collins.

STEM Research Notebook Entry #4

After the read-aloud, ask students to reflect on what they learned about the causes of thunderstorms, tornadoes, and hurricanes in their STEM Research Notebooks. Students should also document how these weather events are alike and different using both words and pictures.

Begin a class vocabulary chart using vocabulary from the interactive read-aloud. Use pictures to illustrate the vocabulary, and post the chart on the classroom wall in a location where it can be easily referenced. You will add to this chart throughout the module.

STEM Research Notebook Entry #5

Have students add the vocabulary words and their definitions to their STEM Research Notebooks, using both words and pictures.

Next, show students a video about tornadoes and their causes, such as "Tornado Facts for Kids!" at *www.youtube.com/watch?v=vH4YT9secVw*. Hold a class discussion about why tornadoes are destructive, emphasizing that winds rotate to form a *vortex* that reaches down from the sky to the ground and moves at very fast speeds. Ask students to share their ideas about what happens to objects in the path of a tornado. Explain that the winds move so fast that objects in a tornado's path can be taken up inside the tornado.

Introduce students to the Vortex Bottle Investigation and the Predict, Observe, Explain process. Students will create STEM Research Notebook entries for each phase of this process. You should also track students' predictions, observations, and explanations on a class chart. Ask students to share their ideas about what a model is, creating a class list. Introduce the idea that a model is a way to represent something that is hard to see or understand in everyday life. Students should understand that models are often smaller representations of larger objects, like model airplanes, but may also be larger representations of things that are too small to see, like the cells that make up the human body. Using students' ideas, formulate a class definition for the term *model* that reflects that a model is a way to show important features of something that is difficult to see in everyday life. Have students brainstorm types of models they have experienced (e.g., model cars, models of the solar system, models of volcanoes erupting). Explain to students that these are called *physical models*, or models that are smaller or larger versions of something else. Have students brainstorm ideas about the benefits of using models (e.g., lets us see and touch things we normally can't). Next, ask students to share their ideas of ways that models are like and unlike the actual object, recording their ideas on a T-chart. Tell students that they are going to create models of the way the wind moves in a tornado. Ask students to brainstorm ideas about how they could do this. After students have shared their ideas, tell them that they will create conditions similar to tornado winds using water in a bottle in the Vortex Bottle Investigation. Tell students that they will make the water move in the same way that tornado winds move. Ask students to describe how tornado winds move, reminding them to use the term *vortex*.

Tell students that they will work in pairs to create a vortex in a bottle and discover what happens to objects in the vortex. Each pair should have all the supplies needed to build a vortex bottle at their table or area. First, however, hold a class discussion, having students predict what will happen by answering the following questions:

- What is a vortex?

- What makes a vortex in nature?

- What are the effects of a wind vortex like a tornado on the environment?

- What will happen to debris in your vortex bottle when you create the vortex?

- What will happen to debris in your vortex bottle after the vortex stops?

Document their predictions on a Predict, Observe, Explain (POE) chart.

STEM Research Notebook Entry #6

Have students write their predictions of what will occur in their vortex bottles during and after the vortex in their STEM Research Notebooks.

After students have completed their predictions, have student pairs conduct the Vortex Bottle Investigation.

Vortex Bottle Investigation

Direct students to do the following:

1. Fill a clear 2 liter plastic bottle about ¾ full with water.

2. Clean up any spilled water with the paper towels.

3. Add a couple of drops of dishwashing liquid.

4. Add a small bit of "debris" (glitter).

5. Put the cap on the bottle tightly.

6. Turn the bottle upside down and spin the bottle in a circular motion.

7. Continue spinning the bottle for about 30 seconds and make observations.

8. Sketch the vortex and the position of the glitter in the chart provided in STEM Research Notebook Entry #7.

9. Allow the water to stop spinning.

10. Sketch the water and the position of the glitter in STEM Research Notebook Entry #7.

Ask students to share their observations of what happened to the debris in the vortex bottle. Document observations on the POE chart.

STEM Research Notebook Entry #7

Students should record their vortex bottle observations, using both words and pictures, as indicated in the previous steps.

Mathematics and Social Studies Connections: Ask students to list ideas of things related to natural hazards that numbers could describe (e.g., inches of rain, speed of wind, number of people affected, number of homes without power, number of hours or days the natural hazard was active). Next, conduct an interactive read-aloud of pages 4–21 of *Hurricanes (Real World Math: Natural Disasters series)*, by Barbara A. Somervill.

STEM Research Notebook Entry #8

After reading the book, have students document what they learned about how they can use mathematics to understand hurricanes in their STEM Research Notebooks, using both words and pictures.

Explanation

Science Class: Have students explain their observations from the Vortex Bottle Investigation first as a class, and then by recording their explanations in their STEM Research Notebooks. As a class, revisit student predictions about the vortex bottles, and ask students if their predictions about what would happen to the glitter were accurate. (They likely observed that the glitter was drawn from the surface of the water into the vortex as it was spinning, and then after the water settled, the glitter sank to the bottom.) Hold a class discussion about how the vortex held the glitter in a compact area while it was spinning. Ask students to consider how objects that are lifted up in a tornado wind vortex would behave. Introduce the idea that when they swirled their bottles, an empty space was created in the center of the water that drew down the surface water with the glitter in it. Ask students if they know what force pulled their glitter downward, and introduce the term *gravity*.

STEM Research Notebook Entry #9

Have students record their explanations of the motion of the glitter in their STEM Research Notebooks.

Remind students that the vortex bottles were models of how winds move in a tornado and that models are not exactly like the thing that they represent. Ask students to share their ideas about how the vortex bottles were like and unlike the winds in a tornado, recording student ideas on a T-chart. Next, ask students to share their ideas about how they could make their models more like an actual tornado, recording students' ideas on chart paper.

Mathematics Connection: Ask students to share their ideas about how they know what the weather will be like tomorrow. Introduce the idea that weather reports they see in the news are based on a science called *meteorology* that uses mathematics to express the chance that the weather will behave in a certain way. Remind students of the discussion they had about models, and introduce the idea that models can also use numbers and equations to show how things work and to make predictions. Introduce the term *mathematical model*, and tell students that meteorologists use mathematical models to predict how the weather will behave. They formulate equations based on conditions in the atmosphere and how the weather has behaved in the past and use these to predict future weather.

Show students a local weather report that includes probabilities of precipitation. Ask them to identify the days when precipitation is most likely and least likely and how they know this. Explain to students that scientists study weather patterns such as air movements and air temperatures to determine whether the conditions are present for rain, snow, or other weather events. Ask students if the weather report is always correct (no). Explain that this is because meteorologists can make predictions based only on the information they have and on how they know the weather has behaved in the past. Since they cannot be sure that a weather event will happen, they use percentages to express the chance, or *probability,* of a weather event occurring. Students should understand that the higher the percentage, the more confident the meteorologist feels about his or her prediction.

Ask students to share what they know and what they want to know about weather reporting and probabilities, recording student ideas on a KLEWS chart. Then, conduct an interactive read-aloud of *That's a Possibility!: A Book About What Might Happen*, by Bruce Goldstone.

STEM Research Notebook Entry #10

After reading the book, have students document what they learned about probabilities in their STEM Research Notebooks, using both words and pictures.

ELA Connection: Remind students that some of the natural hazards they have discussed are related to the weather, including precipitation, such as rain and snow, that falls from the sky. Tell students that they are going to read together a story about a place where food, instead of rain and snow, fall from the sky. Introduce the concept of a tall tale as a story that includes elements that are unbelievable and impossible but that are told as if they could actually happen. Introduce *Cloudy With a Chance of Meatballs*, by Judi Barrett, as a tall tale, and ask students to listen for evidence of a natural hazard as you read the book. Then, conduct an interactive read-aloud. After the read-aloud, ask students to share their ideas about what the natural hazard was in the story (too much food falling from the sky) and how this event is like and unlike an actual natural hazard such as a heavy rainstorm or blizzard. Record student responses on a class T-chart.

Introduce the idea to students that stories have different parts. Have students offer their ideas about what the parts of a story are, creating a class list. Guide students to understand that all stories have some basic parts: beginning, middle, end, characters, setting, and what the story is about (plot). Next, create a two-column chart with columns labeled "Basic Parts of Stories" and "Parts of the Story in *Cloudy With a Chance of Meatballs*." Create a row for each basic story part, and work together as a class to complete the chart.

Social Studies Connection: Remind students of the earlier class discussion when they decided that different kinds of natural hazards were more likely to happen in different places (see p. 59). Point out the places on the map or globe where hazards such as tornadoes, earthquakes, and volcanoes are more likely to happen. Hold a class discussion about being prepared for natural hazards, asking students questions such as the following:

- How would life be the same and different in regions that have more natural hazards compared with regions that do not?

- What are the pros and cons of living in different regions that have different natural hazards?

- How do communities prepare for natural hazards?

- What financial costs do you think natural hazards have for local communities?

Document students' ideas, thoughts, and responses on a KLEWS chart.

Elaboration/Application of Knowledge

Science Class and Social Studies Connection: Hold a class discussion about the possible effects of large amounts of rain, asking students for their ideas. Create a class list. Introduce the idea that a flood occurs when an unusually large amount of rain falls and collects in low-lying area or causes rivers and creeks to overflow. Have students share their personal experiences with large amounts of rain or floods. Ask students what they know about floods and how people can be prepared for a flood and what they wonder about these topics, recording student responses on a KLEWS chart. Conduct an

interactive read-aloud of *Flood Warning (Let's-Read-and-Find-Out Science 2)*, by Katharine Kenah. After the reading, ask students what they learned about floods and preparing for floods, adding to the KLEWS chart. Next, ask students how they learned about floods and flood preparation, prompting students to understand that they learned based on information provided in the book. Introduce the term *evidence* as knowledge that we use to support our beliefs. As a class, review the book to find points in the text that were used as evidence to add to the "E" column of the KLEWS chart.

    Assess student learning in this lesson by asking students to do the following:

- Identify four natural hazards, using words and pictures

- Identify the causes of these four natural hazards

- Choose two of the natural hazards and describe two ways each of these natural hazards could affect people in the community where they happen

- Identify two types of models and give examples of each, using words and pictures

ELA Connection: Have students write and illustrate their own tall tales related to the weather by responding to the following prompt:

> *I woke up one Saturday morning and heard wind outside. I looked out my window and saw …*

Distribute the Creative Writing Rubric (p. 163) and review it with the class. Hand out blank sheets of paper, and have students brainstorm their story ideas using both words and pictures. Once each student has decided on an idea, have students complete the chart provided in STEM Research Notebook Entry #11 to plan their stories.

STEM Research Notebook Entry #11

Have students plan their Weather Tall Tales by completing the chart with the basic parts of the story and the corresponding parts of their own stories. In the "Basic Parts of a Story" column, be sure that students list the following:

- Setting

- Characters

- Plot

- Beginning

- Middle

- End

After students have created their plans, give each student three sheets of lined paper on which to write their stories. Then, after students have completed writing their stories, give each student two sheets of plain white paper on which to draw pictures for their stories. When they are done, have them assemble and staple the pages to create their own books. Have students share their stories with a parent, caregiver, sibling, or friend.

Mathematics Connection: Hand out the narrative you created for the current or recent natural hazard (see Preparation for Lesson 1, p. 58). Have students read the narrative, and ask them to share their observations about how numbers were used. Work as a class to decide how the event could be described using numbers. Introduce the term *data* to the class, and tell them that often numerical data are represented in tables, charts, or graphs so that the data are easier to understand. Create a class table listing relevant items in one column (e.g., inches of rain, number of homes without power, number of people evacuated, lost wages) and the corresponding numerical data in the other. Ask students to share their ideas about whether this is easier or more difficult to understand than the narrative they read, whether there is important information in the narrative that is not included in the table, and how the narrative and the table could be used together to help people understand the impacts of the natural hazard.

Evaluation/Assessment

Students may be assessed on the following performance tasks and other measures listed.

Performance Tasks

- Vortex Bottle Investigation

- Lesson assessment

- Weather Tall Tale (see Creative Writing Rubric, p. 163)

Other Measures (see rubric on p. 162)

- Teacher observations

- STEM Research Notebook entries

- Participation in teams during investigations

INTERNET RESOURCES

Formative assessment
- *http://stemteachingtools.org/pd/sessionc*

Natural hazard information
- *www.usgs.gov/news/earthword-hazard*

- *www.earthtimes.org/encyclopaedia/environmental-issues/natural-disasters*
- *www.usgs.gov/science/science-explorer/Natural+Hazards*
- *https://earthquake.usgs.gov*
- *www.usgs.gov/natural-hazards/landslide-hazards*
- *https://volcanoes.usgs.gov/index.html*
- *www.nasa.gov/content/esd-natural-disasters*

Locations of natural hazards
- *www.redcross.org/get-help/how-to-prepare-for-emergencies/common-natural-disasters-across-us.html*

Tornadoes
- *www.ready.gov/tornadoes*
- *www.kids.nationalgeographic.com/explore/science/tornado/#tornado.jpg*
- *www.weatherwizkids.com/weather-tornado.htm*
- *www.nssl.noaa.gov/education/svrwx101/tornadoes/forecasting*

LEGO Education WeDo resources
- *https://education.lego.com/en-us/lessons/wedo-2-science/robust-structures*
- *https://education.lego.com/en-us/lessons/wedo-2-science/prevent-flooding*
- *https://education.lego.com/en-us/lessons/wedo-2-computational-thinking/volcano-alert*
- *https://education.lego.com/en-us/lessons/wedo-2-science/drop-and-rescue*
- *https://education.lego.com/en-us/lessons/wedo-2-science/hazard-alarm*

Career information
- *www.bls.gov/ooh/home.htm*

KLEWS charts
- *www.nsta.org/publications/news/story.aspx?id=51519*
- *www.nsta.org/store/product_detail.aspx?id=10.2505/4/sc15_052_06_66*

Interactive read-alouds
- *www.readingrockets.org/article/repeated-interactive-read-alouds-preschool-and-kindergarten*

- *www.k5chalkbox.com/interactive-read-aloud.html*

- *www.readwritethink.org/professional-development/strategy-guides/teacher-read-aloud-that-30799.html*

"Natural Hazards" video
- *www.youtube.com/watch?v=_smJ13x90oM*

"Tornado Facts for Kids!" video
- *www.youtube.com/watch?v=vH4YT9secVw*

Lesson Plan 2: Natural Hazard Quest!

In this lesson, students continue to explore natural hazards and connect the impacts of natural hazards with effects on animals' homes. Students investigate earthquakes as an example of a natural hazard caused by phenomena within Earth. Working in teams, students use the engineering design process (EDP) to design and build structures to withstand a simulated earthquake. They conduct research on a natural hazard and present their findings to the class in the Hazard Sleuths activity.

ESSENTIAL QUESTIONS

- What conditions are necessary for natural hazards to occur?

- What influence can natural hazards have on animals' homes?

- What impacts do natural hazards have on people and communities?

ESTABLISHED GOALS AND OBJECTIVES

At the conclusion of this lesson, students will be able to do the following:

- Explain that the movement of tectonic plates can cause natural hazards

- Identify natural hazards associated with movements of tectonic plates

- Identify several impacts that natural hazards associated with the movement of tectonic plates can have on people and communities

- Understand that design features of structures can help protect people during natural hazard events and apply that understanding to create structures designed to withstand a simulated earthquake

- Identify several ways that people can remain safe during a natural hazard occurrence

- Identify the impacts natural hazards can have on animals, with an emphasis on animals' homes

- Identify the steps of the EDP

- Use the EDP to complete a group task

- Communicate information about natural hazards in a visual format

- Use technology to gather research information and communicate about natural hazards

- Use bar graphs to model earthquake data and identify geographic patterns

TIME REQUIRED

- 10 days (approximately 30 minutes each day; see Tables 3.7–3.9, pp. 38–39)

MATERIALS

Required Materials for Lesson 2

- STEM Research Notebooks
- Computer with internet access for viewing videos
- Books:
  - *Earthquakes,* by Ellen Prager (National Geographic Children's Books, 2007)
  - *A House Is a House for Me,* by Mary Ann Hoberman (Puffin Books, 2007)
  - *Lemonade for Sale,* by Stuart J. Murphy (HarperCollins, 1997)
- Chart paper
- Markers
- World map or globe
- Colored pencils (1 set per student)
- 2 sheets of plain white paper (per student)
- 3 sheets of lined writing paper (per student)
- Safety glasses with side shields or safety goggles and nonlatex aprons (per student)

Additional Materials for Earthquake Shake (1 set per team of 2–3 students)

- 1 rectangular baking sheet or pan at least 1 inch deep
- Sand sufficient to cover the bottom of the baking sheet or pan to a depth of ½ inch
- 15 wooden blocks of various sizes (sufficient to create a 9-inch-tall tower)
- 5 wooden sticks (e.g., paint stirrers or large craft sticks)
- Set of plastic interlocking blocks (sufficient to create a 9-inch-tall tower)
- 10 plastic containers or cups (1 oz. size)
- 25 index cards (3 × 5 inches)
- 5 rubber bands (nonlatex)
- Ruler

Additional Materials for Hazard Sleuths (1 per team of 2–3 students unless otherwise indicated)

- Poster board
- Glue
- Set of markers

SAFETY NOTES

1. All students must wear sanitized safety glasses with side shields or safety goggles and nonlatex aprons during all phases of this inquiry activity.

2. Use caution in working with sharps (if the sticks don't have rounded ends), as these can cut or puncture skin.

3. Immediately clean up any sand spills on the floor to avoid a slip-and-fall hazard.

4. Wash hands with soap and water after completing the cleanup phase of the activity.

CONTENT STANDARDS AND KEY VOCABULARY

Table 4.4 lists the content standards from the *NGSS, CCSS,* NAEYC, and the Framework for 21st Century Learning that this lesson addresses, and Table 4.5 (p. 76) presents the key vocabulary. Vocabulary terms are provided for both teacher and student use. Teachers may choose to introduce some or all of the terms to students.

Table 4.4. Content Standards Addressed in STEM Road Map Module Lesson 2

NEXT GENERATION SCIENCE STANDARDS
PERFORMANCE EXPECTATIONS
• 2-PS1-2. Analyze data obtained from testing different materials to determine which materials have the properties that are best suited for an intended purpose.
• 2-PS1-3. Make observations to construct an evidence-based account of how an object made of a small set of pieces can be disassembled and made into a new object.
• 2-LS4-1. Make observations of plants and animals to compare the diversity of life in different habitats.

Continued

Table 4.4. (*continued*)

- 2-ESS1-1. Use information from several sources to provide evidence that Earth events can occur quickly or slowly.

- ETS1-1. Ask questions, make observations, and gather information about a situation people want to change to define a simple problem that can be solved through the development of a new or improved object or tool.

- ETS1-2. Develop a simple sketch, drawing, or physical model to illustrate how the shape of an object helps it function as needed to solve a given problem.

- ETS1-3. Analyze data from tests of two objects designed to solve the same problem to compare the strengths and weaknesses of how each performs.

SCIENCE AND ENGINEERING PRACTICES

Analyzing and Interpreting Data

Analyzing data in K–2 builds on prior experiences and progresses to collecting, recording, and sharing observations.
- Use observations (firsthand or from media) to describe patterns in the natural world in order to answer scientific questions.

Developing and Using Models

Modeling in K–2 builds on prior experiences and progresses to include using and developing models (i.e., diagram, drawing, physical replica, diorama, dramatization, storyboard) that represent concrete events or design solutions.
- Use a model to represent relationships in the natural world.

Planning and Carrying Out Investigations

Planning and carrying out investigations to answer questions or test solutions to problems in K–2 builds on prior experiences and progresses to simple investigations, based on fair tests, which provide data to support explanations or design solutions.
- Make observations (firsthand or from media) to collect data that can be used to make comparisons.

DISCIPLINARY CORE IDEAS

LS1.C. Organization for Matter and Energy Flow in Organisms

- All animals need food in order to live and grow. They obtain their food from plants or from other animals. Plants need water and light to live and grow.

ESS3.A. Natural Resources

- Living things need water, air, and resources from the land, and they live in places that have the things they need. Humans use natural resources for everything they do.

Continued

Table 4.4. (*continued*)

ESS2.D. Weather and Climate
- Weather is the combination of sunlight, wind, snow or rain, and temperature in a particular region at a particular time. People measure these conditions to describe and record the weather and to notice patterns over time.

CROSSCUTTING CONCEPTS

Patterns
- Patterns in the natural and human designed world can be observed, used to describe phenomena, and used as evidence.

Systems and System Models
- Systems in the natural and designed world have parts that work together.

Cause and Effect
- Events have causes that generate observable patterns.

COMMON CORE STATE STANDARDS FOR MATHEMATICS

MATHEMATICAL PRACTICES
- MP1. Make sense of problems and persevere in solving them.
- MP2. Reason abstractly and quantitatively.
- MP3. Construct viable arguments and critique the reasoning of others.
- MP4. Model with mathematics.
- MP5. Use appropriate tools strategically.
- MP6. Attend to precision.
- MP7. Look for and make use of structure.
- MP8. Look for and express regularity in repeated reasoning.

MATHEMATICAL CONTENT
- 2.NBT.A.1. Understand that the three digits of a three-digit number represent amounts of hundreds, tens, and ones; e.g., 706 equals 7 hundreds, 0 tens, and 6 ones.
- 2.NBT.A.2. Count within 1,000; skip-count by 5s, 10s, and 100s.
- 2.NBT.A.3. Read and write numbers to 1,000 using base-ten numerals, number names, and expanded form.

Continued

Table 4.4. (*continued*)

COMMON CORE STATE STANDARDS FOR ENGLISH LANGUAGE ARTS

READING STANDARDS

- RI.2.1. Ask and answer such questions as *who, what, where, when, why,* and *how* to demonstrate understanding of key details in a text.

- RI.2.3. Describe the connection between a series of historical events, scientific ideas or concepts, or steps in technical procedures in a text.

- RI.2.7. Explain how specific images (e.g., a diagram showing how a machine works) contribute to and clarify a text.

- RI.2.8. Describe how reasons support specific points the author makes in a text.

- RI.2.9. Compare and contrast the most important points presented by two texts on the same topic.

WRITING STANDARDS

- W.2.1. Write opinion pieces in which they introduce the topic or book they are writing about, state an opinion, supply reasons that support the opinion, use linking words (e.g., *because, and, also*) to connect opinion and reasons, and provide a concluding statement or section.

- W.2.2. Write informative/explanatory texts in which they introduce a topic, use facts and definitions to develop points, and provide a concluding statement or section.

- W.2.6. With guidance and support from adults, use a variety of digital tools to produce and publish writing, including in collaboration with peers.

- W.2.7. Participate in shared research and writing projects (e.g., read a number of books on a single topic to produce a report; record science observations).

- W.2.8. Recall information from experiences or gather information from provided sources to answer a question.

SPEAKING AND LISTENING STANDARDS

- SL.2.1. Participate in collaborative conversations with diverse partners about *grade 2 topics and texts* with peers and adults in small and larger groups.

- SL.2.2. Recount or describe key ideas or details from a text read aloud or information presented orally or through other media.

- SL.2.3. Ask and answer questions about what a speaker says in order to clarify comprehension, gather additional information, or deepen understanding of a topic or issue.

Continued

Table 4.4. (*continued*)

NATIONAL ASSOCIATION FOR THE EDUCATION OF YOUNG CHILDREN STANDARDS
• 2.G.02. Children are provided varied opportunities and materials to learn key content and principles of science.
• 2.G.03. Children are provided varied opportunities and materials that encourage them to use the five senses to observe, explore, and experiment with scientific phenomena.
• 2.G.04. Children are provided varied opportunities to use simple tools to observe objects and scientific phenomena.
• 2.G.05. Children are provided varied opportunities and materials to collect data and to represent and document their findings (e.g., through drawing or graphing).
• 2.G.06. Children are provided varied opportunities and materials that encourage them to think, question, and reason about observed and inferred phenomena.
• 2.G.07. Children are provided varied opportunities and materials that encourage them to discuss scientific concepts in everyday conversation.
• 2.G.08. Children are provided varied opportunities and materials that help them learn and use scientific terminology and vocabulary associated with the content areas.
• 2.H.02. All children have opportunities to access technology (e.g., tape recorders, microscopes, computers) that they can use.
• 2.H.03. Technology is used to extend learning within the classroom and integrate and enrich the curriculum.
FRAMEWORK FOR 21ST CENTURY LEARNING
• Interdisciplinary Themes; Learning and Innovation Skills; Information, Media, and Technology Skills; Life and Career Skills.

Table 4.5. Key Vocabulary for Lesson 2

Key Vocabulary	Definition
collaboration	the act of working together as a team to achieve a goal
crust (of Earth)	the outermost layer of Earth's surface
erupt	to send out materials such as lava and hot gases
lava	very hot liquid material that comes out of the mantle of Earth in a volcanic eruption; magma that reaches Earth's surface
magma	melted rock that is below Earth's surface; expelled from volcanoes as lava

Continued

Table 4.5. (*continued*)

Key Vocabulary	Definition
magnitude	the size of something; for earthquakes, the energy released where the tectonic plates collide
mantle (of Earth)	the layer of Earth beneath the crust made of rocks and hot liquid; makes up most of the inside of Earth
natural resources	something in nature that can be used by people
seismic waves	waves of energy that travel through Earth's surface because of an earthquake
subduction zone	region of Earth where tectonic plates collide
tectonic plate	very large slabs of rock within Earth's mantle
tephra	rocks ejected from a volcano during an eruption
tremor	a very small earthquake
tsunami	a very large ocean wave often caused by an earthquake that occurs at the floor of the ocean

TEACHER BACKGROUND INFORMATION

Whereas Lesson 1's activities focused on natural hazards associated with atmospheric conditions, this lesson has students investigates natural hazards that are caused by phenomena within Earth. Students use the EDP and their learning about earthquakes to design a solution to a problem.

Earthquakes and Associated Natural Hazards

Earthquakes occur when pieces of Earth's crust, called *tectonic plates*, collide with one another. These movements release energy that can sometimes be felt in the shaking of Earth's surface. It is important to note that earthquakes are relatively frequent, with about 50 per day occurring around the world. Many of these produce only small amounts of energy, so the tremors they produce are imperceptible, and many occur in areas not inhabited by humans or at the bottom of the ocean. Geologists are unable to predict earthquakes, although they can forecast probabilities of earthquakes occurring in a given area (see *www.usgs.gov/faqs/are-earthquake-probabilities-or-forecasts-same-prediction?qt-news_science_products=7#qt-news_science_products* for more information).

Earthquakes are measured by the magnitude of the vibrations, or seismic waves, produced when pieces of Earth's crust collide and move against one another. Magnitudes are measured on the moment magnitude scale, expressed numerically on a scale from 1 to 10,

where the magnitude is associated with the potential for damage to human life and property. On this scale, for example, a magnitude 7.0–7.9 earthquake would be considered a major earthquake, producing serious damage (see *www.geo.mtu.edu/UPSeis/magnitude.html* for more information). Earthquakes are associated with other natural hazards, including tsunamis and volcanoes. See the following websites for additional information about earthquakes and Earth's geological makeup:

- *https://earthquake.usgs.gov*

- *www.ready.gov/earthquakes*

- *https://disasters.nasa.gov/earthquakes*

- *www.kids.nationalgeographic.com/explore/science/earthquake/#earthquake-houses.jpg*

Tsunami is a Japanese word that means "harbor wave." Several events, including landslides, volcanic eruptions, and meteorites, can cause tsunamis, but most tsunamis are caused by earthquakes that occur at the floor of the ocean and result in a series of seismic waves. These waves may appear to be relatively small in the open ocean but can move as fast as 600 mph. As the waves approach the shore, they tend to slow down and can grow much larger. Because tsunamis are not dependent on weather patterns or tides, they can occur at any time of the year. Tsunamis are difficult to detect, since the waves they create in the open ocean are relatively small. Detecting earthquakes at the ocean floor presents logistical and technological challenges, although new approaches to detection, involving technologies such as GPS and ionosphere observation (see *www.nasa.gov/feature/jpl/scientists-look-to-skies-to-improve-tsunami-detection* for more information), are being developed in response to recent tsunamis that resulted in high mortality, such as the one that struck Indonesia in 2018. The difficulty of detecting tsunamis and their rapid movement mean that they can pose significant threats to coastal communities. Warning signs of tsunamis include earthquakes felt on land in coastal areas, a loud roar from the ocean, and unusual tidal behavior, including a sudden rise in water or outflow of water from the coast that exposes the ocean floor. Tsunamis tend to happen in certain geographic locations because of the geological composition of those areas. In particular, tsunamis are most common around the Pacific Ocean basin, an area commonly called the "Ring of Fire." This area contains a number of subduction zones, or areas where tectonic plates collide. Additional information about tsunamis can be found on the following websites:

- *www.ready.gov/tsunamis*

- *www.natgeokids.com/uk/discover/geography/physical-geography/tsunamis*

- *http://tsunami.org/what-causes-a-tsunami*

- *http://news.bbc.co.uk/2/hi/asia-pacific/135797.stm*

The word *volcano* is derived from the name of the Roman god of fire, Vulcan. A volcano is an opening in Earth, often in a mountain, that allows underground material that is warmer than the surrounding environment to escape. The escape of these materials—gases, hot liquid rock called *magma* (referred to as *lava* after it escapes), and solid pieces of rock—is called an *eruption*. Eruptions can be classified as explosive, in which magma is torn apart as it quickly rises, resulting in matter being shot violently into the air, or effusive, in which magma rises more slowly, resulting in lava flowing out of cracks and other openings. The 1980 eruption of Mount St. Helens was explosive (see *https://volcanoes.usgs.gov/volcanoes/st_helens* for more information), whereas the long-running eruption of Kīlauea Volcano on the Island of Hawaii has been primarily effusive (see *https://pubs.usgs.gov/fs/2013/3116* for more information).

Volcano eruptions occur when magma rises because of its relatively low density compared with the surrounding rock. As the magma rises, gases dissolved in the magma form bubbles that create pressure. This pressure causes the magma to rise to Earth's surface, where it escapes through openings, sometimes violently. Lava flows at a relatively slow pace and is therefore not the primary threat to humans associated with volcanic eruptions. More dangerous are the broken pieces of magma called *tephra*, toxic gases, and large amounts of ash that can be released. In addition, volcanoes can cause mudflows when expelled materials mix with water in streams or with snow or ice; these mudflows can be dangerous to people and property. Volcanoes are considered dormant if they have not erupted in about 10,000 years, active if they have erupted within that time, or extinct if they have low levels of lava and have not erupted within the last 10,000 years. However, there are recorded instances of eruptions from volcanoes that were considered extinct, such as Colli Albani, a volcano near Rome that began showing signs of activity in 2016 (see *www.livescience.com/55397-extinct-rome-volcano-rumbles-to-life.html* for more information). For additional information about volcanoes, see the following websites:

- *https://volcanoes.usgs.gov/index.html*
- *www.ready.gov/volcanoes*
- *www.spaceplace.nasa.gov/volcanoes2/en*
- *www.natgeokids.com/uk/discover/geography/physical-geography/volcano-facts*
- *www.geology.com/volcanoes/types-of-volcanic-eruptions*

Engineering

Students begin to gain an understanding of engineering as a profession in this lesson as they learn to use the EDP to create earthquake-proof structures. Students should understand that engineers are people who design and build products and systems in response

to human needs. For an overview of the various types of engineering professions, see the following websites:

- *www.engineergirl.org/33/TryOnACareer*

- *www.nacme.org/types-of-engineering*

- *www.sciencekids.co.nz/sciencefacts/engineering/typesofengineeringjobs.html*

In this lesson, students work in teams to create structures that can withstand shaking that simulates earthquake tremors. Students use simple sets of materials such as wooden or interlocking plastic blocks to explore various types of structural designs. Earthquake-resistant architecture and design is a dynamic field; however, the advanced physics principles associated with these fields are beyond the scope of this module. Generally speaking, symmetrical buildings are more earthquake resistant than asymmetrically designed structures (for example, L-shaped buildings). More information about earthquake-resistant structures is available at the following websites:

- *www.imaginationstationtoledo.org/educator/activities/can-you-build-an-earthquake-proof-building*

- *https://science.howstuffworks.com/engineering/structural/earthquake-resistant-buildings3.htm*

Engineering Design Process

Students should understand that engineers need to work in groups to accomplish their work, and that collaboration is important for designing solutions to problems. Students will use the EDP, the same process that professional engineers use in their work, in this lesson. A graphic representation of the EDP is provided at the end of this chapter (p. 114). You may wish to provide each student with a copy of the EDP graphic or enlarge it and post it in a prominent place in your classroom for student reference throughout the module. Be prepared to review each step of the EDP with students, and emphasize that the process is not a linear one—at any point in the process, they may need to return to a previous step. The steps of the process are as follows:

1. *Define.* Describe the problem you are trying to solve, identify what materials you are able to use, and decide how much time and help you have to solve the problem.

2. *Learn.* Brainstorm solutions and conduct research to learn about the problem you are trying to solve.

3. *Plan.* Plan your work, including making sketches and dividing tasks among team members if necessary.

4. *Try.* Build a device, create a system, or complete a product.

5. *Test.* Now, test your solution. This might be done by conducting a performance test, if you have created a device to accomplish a task, or by asking for feedback from others about their solutions to the same problem.

6. *Decide.* Based on what you found out during the Test step, you can adjust your solution or make changes to your device.

After completing all six steps, students can share their solution or device with others. This represents an additional opportunity to receive feedback and make modifications based on that feedback.

The following are additional resources about the EDP:

- *www.sciencebuddies.org/engineering-design-process/engineering-design-compare-scientific-method.shtml*

- *www.pbslearningmedia.org/resource/phy03.sci.engin.design.desprocess/what-is-the-design-process*

COMMON MISCONCEPTIONS

Students will have various types of prior knowledge about the concepts introduced in this lesson. Table 4.6 (p. 82) outlines some common misconceptions students may have concerning these concepts. Because of the breadth of students' experiences, it is not possible to anticipate every misconception that students may bring as they approach this lesson. Incorrect or inaccurate prior understanding of concepts can influence student learning in the future, however, so it is important to be alert to misconceptions such as those presented in the table.

Table 4.6. Common Misconceptions About the Concepts in Lesson 2

Topic	Student Misconception	Explanation
Engineers and the engineering design process (EDP)	Engineers are people who drive trains.	Railroad engineers are just one type of engineer. The engineers referred to in this module are people who use science, technology, and mathematics to build machines, products, and structures that meet people's needs.
	Engineers use only science and mathematics to do their work.	Engineers often use science and mathematics in their work, but they also use many other kinds of knowledge to solve problems and design products, such as how people use products, what people's needs are, and how the natural environment affects materials.
	Engineers work alone to build things.	Engineers often work in teams and use a process to solve problems. The process involves creative thinking, research, and planning, in addition to building and testing products.
Natural hazards	Earthquakes can occur only along fault lines.	Earthquakes can occur anywhere on land or in the water.
	The ground opens up along a fault line during an earthquake.	While cracks in the ground can form, large gaps are not created along fault lines, since the earthquake is caused by the plates of Earth's crust moving past each other. If they moved completely apart to form a large opening, there would be no earthquake.
	Earthquakes always cause a great deal of damage.	Only a small percentage (about 20%) of earthquakes that occur are even felt by people.
	If a volcano has not erupted in many years, it is extinct.	Volcanoes can remain dormant for many years without erupting; they are considered extinct only when they no longer have a lava supply, but this is difficult to determine.
	Volcanoes always erupt violently, like an explosion.	The way a volcano erupts depends on the dissolved gases and other matter in the volcano. Some volcanoes erupt by magma and gas seeping out from cracks in the sides rather than erupting violently.
	An erupting volcano is dangerous only because of the hot lava that flows from it.	While lava flows are dangerous to things in their path, volcanoes also release gases and ash that can be dangerous.

PREPARATION FOR LESSON 2

Review the Teacher Background Information section (p. 77), assemble the materials for the lesson, and duplicate the EDP graphic (p. 114) if you wish to hand it out to students or enlarge it to post in the classroom.

Identify and have available several before-and-after pictures that show the impacts of natural hazards, including both natural hazard events related to the weather (e.g., floods, hurricanes, tornadoes) and natural hazard events related to the movement of plates in Earth's crust (e.g., earthquakes, volcanoes, tsunamis). You can find these by conducting an internet search for images using search terms such as "before-and-after pictures of natural disasters."

For the current or recent natural disaster you identified in Lesson 1, create a written narrative about the event appropriate to your students' reading level that emphasizes the financial costs associated with the disaster. This may include estimated costs of the impacts on natural resources (e.g., lost crops), costs associated with correcting damage to natural resources (e.g., beach erosion, replanting forests, repairing public parks), and other costs such as repairing roads, lost wages from business closures, and lost tourist revenue. If this information is not readily available in news reports, the NOAA provides information on the costs of natural disasters at *www.ncei.noaa.gov/news/calculating-cost-weather-and-climate-disasters*.

In this lesson, student teams will create earthquake-resistant structures using simple sets of materials. To test their structures, teams will need to simulate earthquakes by either shaking the containers in which they built their structures or using an earthquake shake table you provide. If your school does not have an earthquake shake table, you can create one from simple materials such as cardboard, rubber bands, tennis balls, masking tape, and paint stirrers using the directions provided at *www.pbskids.org/designsquad/build/seismic-shake-up*.

Each student team will conduct research on a natural hazard and create a poster that features what they learned about the hazard, its causes, its effects on people and the environment, and how people can stay safe if this hazard should occur in their community. You should either bookmark websites on a computer with internet access for each team or print out appropriate information for hurricanes, tornadoes, earthquakes, tsunamis, and volcanoes. Note that for the module challenge, each team will create its preparedness plan for the hazard assigned in this lesson, so you may wish to assign natural hazards that are common in your area.

In addition to the websites provided in the Teacher Background Information section, the following provide information on natural hazards and preparedness:

Hurricane
- *www.scied.ucar.edu/webweather/hurricanes*

- *www.fema.gov/media-library-data/253331ac179d32652a4d0cbf7fb3e6eb/FEMA_FS_hurricanes_508_8-15-13.pdf*

Tornado

- *www.scied.ucar.edu/webweather/tornadoes*

- *www.fema.gov/media-library-data/a4ec63524f9fd1fa5d72be63bd6b29cf/FEMA_FS_tornado_508_8-15-13.pdf*

Earthquake

- *https://earthquake.usgs.gov/learn/kids*

- *www.fema.gov/media-library-data/ed51897eb6583e40ec1edb5f8fb85ee5/FEMA_FS_earthquake_508_081513.pdf*

Tsunami

- *www.ready.gov/kids/know-the-facts/tsunamis*

- *www.fema.gov/media-library-data/26ea1ceee22e73558b01030a4a0d7c47/FEMA_FS_tsunami_508-8-15-13.pdf*

Volcano

- *www.ready.gov/kids/know-the-facts/volcano*

- *www.fema.gov/media-library-data/a4402e44902b963c8de7ee4ad0586016/FEMA_FS_volcano_508-8-15-13.pdf*

For the mathematics and social studies connections in this lesson, students will create bar graphs of recent earthquakes' magnitudes, organized by continent. Recent earthquakes with magnitude 2.5 and higher are listed on the USGS web page "Earthquakes" at *https://earthquake.usgs.gov/earthquakes*; click on "Latest Earthquakes." You should be prepared to share this web page with your class. Since students will use data from this site to create their graphs, you should ensure that they have access to the site or, alternatively, print out copies of the map and accompanying data for each student. In addition, you should print out a copy of a map of the U.S. regions for each student (for example, the U.S. Census Bureau map at *www2.census.gov/geo/pdfs/maps-data/maps/reference/us_regdiv.pdf*).

LEARNING COMPONENTS
Introductory Activity/Engagement

Connection to the Challenge: Begin each day of this lesson by directing students' attention to the module challenge, the Natural Hazard Preparedness Challenge:

Your town's leaders want to be sure that people will be safe in case a natural hazard should strike your town. They have asked your class to create preparedness plans for natural hazards that will keep people safe during these events. You and your team are challenged to create a plan for the community for one natural hazard and an advertisement that lets people know about your plan.

Hold a brief class discussion of how students' learning in the previous days' lessons contributed to their ability to complete the challenge. You may wish to create a class list of key ideas on chart paper.

Science Class and ELA Connection: Remind students that in the last lesson they discussed tornadoes and floods. Ask students what the cause of those natural hazards is (weather). Next, remind students that other natural disasters are caused by movements within Earth. Introduce the idea that the part of Earth that we see at the surface is called the *crust,* and that the crust is a relatively thin layer that covers Earth's *mantle.* Tell students that the mantle actually makes up most of the inside of Earth and is made mostly of rock, but that some parts of the mantle are very hot and liquid. Ask students if they can think of an example of when some of the hot liquid substance from Earth's mantle rises to the surface (lava).

Next, introduce the idea that even though we may not feel it, Earth's crust is always moving. This is because Earth's surface is not one continuous solid surface but is made up of large, moving pieces called *tectonic plates* in Earth's crust and mantle. Ask students for their ideas about what might happen when these moving plates run into each other (natural hazards such as earthquakes, volcanoes, and tsunamis). Tell students that they will be investigating earthquakes in this lesson. Create a KLEWS chart for earthquakes and ask students to share what they know about earthquakes, recording their ideas on the chart. Then, conduct an interactive read-aloud of the book *Earthquakes,* by Ellen Prager, documenting student responses on the KLEWS chart.

STEM Research Notebook Entry #12

Have students document what they learned about earthquakes in their STEM Research Notebooks, using both words and pictures.

Next, hold a discussion about the impacts of natural hazards on people and communities, documenting student responses on a KLEWS chart. Following agreed-upon rules for discussions, ask students the following questions:

- What impacts do natural hazards have on people? (e.g., losing power, flooding in homes, destruction of property, injury to people)

- What impacts can natural hazards have on communities? (e.g., roads flooded, debris in roads, beach erosion, trees blown down)

Show students the before-and-after images you found (see p. 83) and hold a class discussion for each picture, including the type of natural hazard that may have occurred, the cause of that hazard (weather causes or movement of Earth's crust), how the natural hazard affected the community in which it occurred, and how the natural hazard affected the natural environment specifically.

Ask students to consider what other living things, besides people, may have been affected in the natural hazards in the pictures. Tell students that all animals and plants live in habitats. Ask students to share their ideas about what a habitat is, recording these on chart paper. Work with students to formulate a class definition of habitat that reflects the idea that a habitat is the environment where people, other animals, and plants live and where their needs, such as shelter, food, and water, are met. Ask students to identify some types of habitats they are familiar with and create a class list (e.g., forest, ocean, river, rain forest, city).

Ask students to think about the places where they get their shelter, food, water, and other life necessities, and encourage them to think more specifically about habitats where animals and people live. Introduce the idea that students' houses are an important part of their habitat. Next, ask students about various animals' homes (e.g., bird, fish, snake, cow, deer, dolphin, turtle), and create a class list of responses. Next, ask students to consider which animals' homes might be affected by a tornado, a flood, and an earthquake, adding their responses to the class list. Then, conduct an interactive read-aloud of *A House Is a House for Me,* by Mary Ann Hoberman, or alternatively, show the video of the author reading the book found at *www.youtube.com/watch?v=s4cGB0-7R-0.* After the interactive read-aloud, discuss with students how the animal homes mentioned in the book provide for the animals' needs. Ask students to identify the animals and animal homes featured in the book, and add these to the class list.

Show students the before-and-after images again, and ask them, for each image, what animal homes might have been affected by the natural hazard.

STEM Research Notebook Entry #13

Choose three of the animals and homes on the class list, and have students document in their STEM Research Notebooks their ideas of how a tornado, a flood, and an earthquake might affect these three animals' homes, using both words and pictures.

Mathematics and Social Studies Connections: Have the class continue to consider the current natural hazard you identified in Lesson 1. Introduce the term *natural resources.* Discuss as a class the impacts a natural hazard could have on natural resources in a community (e.g., it might damage farms that grow food, kill trees and other plants in parks, erode beaches). As a class, work to document the impacts the current natural hazard is having on the local community in terms of natural resources, recording ideas on a class chart.

Next, hand out the narrative you prepared about the financial costs of the natural disaster (see p. 83), and read it as a class. Point out to students that some of the impacts on natural resources will have financial costs associated with them (e.g., lost crops or the costs of replanting trees or correcting beach erosion). Besides these costs, there are many other impacts that natural hazards can have on communities. Have students identify the financial impacts associated with the disaster from the narrative, and record these on a class chart.

STEM Research Notebook Entry #14

Have students record the financial impacts of the current natural disaster.

Activity/Exploration

Science Class: Ask students to identify the potential effects of earthquakes on communities, and record student ideas on chart paper. Tell students that special building techniques are often used for buildings constructed in areas that frequently experience earthquakes to help them withstand the tremors associated with earthquakes and keep the occupants of the buildings safe. Ask students for their ideas about who designs buildings (architects, engineers). Introduce the idea that engineers use a process called the engineering design process (EDP) to do their work. Describe the steps of the EDP using the graphic provided on page 114.

Next, introduce the idea that engineers and other design professionals such as architects often work in groups or teams to do their work, and that they will act as engineers in this lesson to create earthquake-resistant buildings. Introduce the term *collaboration.* Create a T-chart labeled "collaboration," with one column labeled "benefits" and the other labeled "challenges." Ask students to share their ideas about why engineers collaborate (e.g., having many people's ideas to draw on, sharing work, being able to talk about ideas, having people with different skills and strengths working on a project). Then, ask students to think about experiences they have had in working with groups, and have them share the challenges they faced (e.g., too many people trying to talk at once, one person doing all the work, people not listening to each other). Emphasize to students that collaborating requires all team members to participate and be respectful of each other. Group students in teams of two or three to work together for the Earthquake Shake activity. Ask the teams of students to brainstorm some rules for good collaboration that will help their team work together well. Then, have the teams share their ideas to create a class list. Based on students' ideas, create a class "contract" for good collaboration that incorporates students' ideas into several rules they will agree to follow.

STEM Research Notebook Entry #15

Have students record the class collaboration contract in their STEM Research Notebooks and sign their names to it as an agreement that they will follow these rules.

Introduce the Earthquake Shake activity by telling students that they will act as engineers working in teams to build structures that they think can withstand earthquake conditions. Have students work in the teams that were formed earlier. Provide each team with a set of building materials. Review safety notes. Tell the teams that they will need to build 9-inch-tall structures on their baking sheets or pans. Have students spread the sand so that it covers the bottom of the baking sheets or pans to a depth of ½ inch. Model how to use the ruler to measure the depth of the sand.

Tell students that they will build structures using the materials they have been given to create earthquake-resistant "buildings." Review the procedure for the activity, emphasizing to students that they will plan and predict (using the first steps of the EDP) before they build anything. Students are challenged to use the provided materials to design and build a structure at least 9 inches tall that will withstand a simulated earthquake. If you do not have access to an earthquake shake table (see p. 83), demonstrate to students how to shake their pans gently to simulate a mild earthquake and more vigorously to simulate a more powerful earthquake. You may wish to shake the pans for teams' final tests to ensure consistency of the power of the "earthquakes." Work through the steps of the EDP and the Predict, Observe, Explain (POE) cycle by instructing teams to consider the questions listed below. Have teams share their ideas with the class, recording them on a POE chart. Then, have students create STEM Research Notebook entries for each set of steps.

1. Define: What is the problem, and what are you trying to do?

2. Learn (predict): What do you need to think about to solve the problem? What ideas does your team have? Predict what materials will make the building most earthquake resistant.

STEM Research Notebook Entry #16

Students should record their progress in steps 1 and 2 in their STEM Research Notebooks.

3. Plan: Choose the materials you think will create the most earthquake-proof structure. Create a drawing of the structure you will build, labeling your materials.

4. Try: Build the structure you planned.

5. Test (observe): Observe how the structure withstands simulated earthquakes, shaking the pan to simulate minor and more powerful earthquakes.

STEM Research Notebook Entry #17

Students should record their progress in steps 3–5 in their STEM Research Notebooks.

6. Decide (explain). Can you improve your design? Explain why you think your structure behaved the way it did. Try some different designs based on your test.

STEM Research Notebook Entry #18

Students should record their progress in step 6 in their STEM Research Notebooks.

Students may proceed through the Plan, Try, and Test steps of the EDP several times as they create different designs.

Mathematics and Social Studies Connections: Ask students to share their ideas about how earthquake strengths can be measured, creating a class list. Introduce the term *magnitude,* explaining that magnitude is a measurement of the strength of an earthquake. An earthquake's magnitude tells how much energy is released where the earthquake happens (i.e., where plates collide with each other). Tell students that earthquake magnitudes are measured on a scale from 1 to 10, with 1 being an earthquake that can't be felt by people, and 10 being an extremely large earthquake that would cause serious damage. (*Note:* The largest earthquake ever recorded occurred in 1960 in Chile, with a magnitude of 9.5.)

Remind students that earthquakes are most likely to happen at places where the plates under Earth's surface meet, although they can happen anywhere. Using the world map or globe, have students name and point out the seven continents. Ask students for their ideas about where most earthquakes happen. Tell students that earthquakes actually happen every day. Show students the USGS website on current earthquakes (accessed from *https://earthquake.usgs.gov/earthquakes*). Show students the map provided on the website, and have them point out the continents on this map. Next, ask students where they see most earthquakes. Ask students if the map tells them anything about the size of those earthquakes (no, a separate list gives the magnitude, or you must click on the map to learn details about the earthquake). Have students give their ideas about whether the earthquake's location and size could be presented together in a way that is easy to understand. Introduce the idea that graphs allow scientists and others to provide information such as the location and size of earthquakes in a way that is easy to understand. Ask students to recall the discussion the class had about physical models in Lesson 1.

Remind students that things can be modeled with numbers as well as with physical materials and that graphs are a way to model information mathematically. Conduct an interactive read-aloud of *Lemonade for Sale,* by Stuart J. Murphy, to provide an example of bar graphs and how they can be used to illustrate numerical information.

Tell students that they will use the data from the USGS website about current earthquakes to create bar graphs that will show their magnitudes and the continents on which they occurred.

STEM Research Notebook Entry #19

Have each student create a bar graph with the continents on the x-axis and magnitudes (1–10) on the y-axis in a STEM Research Notebook entry. Students should group earthquakes from a single continent together and use a single color to represent earthquakes within a continent. Provide an example for students of how to group bars by continent, outline each bar, and color it in using their colored pencils. Have students include no more than three bars per continent.

ELA Connection: Student teams will conduct research on natural hazards in the Hazard Sleuths activity. Students should work in the same teams of two or three for this activity as they did for the Earthquake Shake activity. Assign each team one natural hazard (e.g., hurricane, tornado, earthquake, tsunami, volcano). Note that each team will create its preparedness plan for the hazard assigned in this lesson, so you may wish to assign natural hazards that are common in your area. Either direct students to appropriate websites (see p. 93) or distribute the information you printed in advance.

STEM Research Notebook Entry #20

Have each student record his or her team's research findings in a STEM Research Notebook entry, including the following information: the name of the natural hazard, times of year and places that it is most likely to occur, the cause of the natural hazard, its impacts on people, and what people can do to stay safe if the natural hazard affects their community.

Explanation

Science Class: Ask each student team to identify its best design (most earthquake resistant) from the Earthquake Shake activity and share this design with the class, explaining what design features the team members think made this design worked best. Hold a class discussion about what features the teams' best designs share and what unique features students notice. Test each team's best design with increasingly vigorous shaking to determine which team's design is the most earthquake resistant.

Revisit students' predictions from before the investigation, and ask them to compare their predictions with the results of shaking the class's most earthquake-resistant building. Emphasize to the class that not only the materials but also the way the materials were used (the design) was important.

Mathematics Connection: Have several students share their bar graphs. (Since there are multiple earthquakes per continent, students may have included different earthquakes.) Based on students' graphs, ask the class to share ideas about where most earthquakes occur. Return to the USGS website on current earthquakes and compare the class's decision about the location of most earthquakes with the data on the map.

Now, use the USGS website to show the class earthquake data worldwide for the past 30 days. This can be accessed through *https://earthquake.usgs.gov/earthquakes/browse*. Click on "Search Earthquake Catalog" at the bottom of the "By Location" section, select "Past 30 Days," and then click on "Search." This will likely show that most earthquakes occur in the area known as the Ring of Fire around the Pacific Ocean basin. Ask students to use what they know about the cause of earthquakes to explain why so many earthquakes occur there (there are many places where tectonic plates meet and collide). Ask students to share their ideas about what other natural hazards may also be likely in this area (volcanoes and tsunamis, since earthquakes are a cause of these hazards).

ELA and Social Studies Connections: Have student teams use the information from the Hazard Sleuths activity to create a poster about the natural hazard they investigated to educate their classmates about the hazard and what people can do to stay safe if the hazard should strike their community. Have students decide on a title for the posters and include that at the top, then tell them to divide their posters into three sections as follows: background information about the hazard (where it happens, why it happens), possible impacts on people and the environment, and how people can stay safe during this hazard.

Elaboration/Application of Knowledge

Science Class: Assess student learning in this lesson by asking students to do the following, using the natural hazard they studied for the Hazard Sleuths activity:

- Describe what causes this natural hazard to occur

- Identify three impacts this natural hazard can have on people and the environment

- Describe two ways people can stay safe during this natural hazard

Mathematics and Social Studies Connections: Remind students that earthquakes are most likely to occur in the Ring of Fire near the Pacific Ocean. Have students look at the

United States on the map or globe, and ask them if this information helps them predict where most earthquakes occurred in the United States during the past month (near the Pacific Ocean). Tell students that they are going to see if this prediction is accurate. Tell students that the United States is made up of different regions. Show students a map of U.S. regions such as the one provided by the U.S. Census Bureau at *www2.census.gov/ geo/pdfs/maps-data/maps/reference/us_regdiv.pdf*. Tell students that they are going to graph recent earthquakes in the United States according to what region they occurred in (West, Midwest, Northeast, or South) using a regional map and earthquake data. Have students use USGS data about earthquakes within the United States in the past 30 days to create bar graphs for earthquakes in the four regions of the United States. (Note: You may wish to have students divide the West region into Pacific and Mountain regions, as shown on the U.S. Census Bureau map, to emphasize that most earthquakes likely occurred near the Pacific coast.) Students can interact with the maps and data online, or alternatively, you may wish to print out the data.

STEM Research Notebook Entry #21

Have students create bar graphs of the magnitudes of current earthquakes organized by U.S. region in their STEM Research Notebooks.

ELA Connection: Have students use the knowledge they have gained about natural hazards to create a story about a natural disaster's impacts on an animal's home. Review the Creative Writing Rubric (p. 163) with the class. Hand out blank sheets of paper, and have students brainstorm their story ideas using both words and pictures.

STEM Research Notebook Entry #22

Once students have decided on their story ideas, have them create plans for their animal stories. Each student should complete the chart in this STEM Research Notebook entry with the basic parts of the story and the corresponding parts of his or her own story. In the "Basic Parts of a Story" column, be sure that students list the following:

- Setting
- Characters
- Plot
- Beginning
- Middle
- End

Next, give each student three sheets of lined paper on which to write their stories. After students have finished writing their stories, give each student two sheets of plain white paper on which to draw pictures for their stories. When students have completed their stories and pictures, have them assemble and staple the pages to create a book. Have students share their stories with a parent, caregiver, sibling, or friend.

Evaluation/Assessment

Students may be assessed on the following performance tasks and other measures listed.

Performance Tasks

- Earthquake Shake structures and group presentations

- Hazard Sleuths research and poster

- "Animals in a Natural Hazard" story (creative writing)

- Lesson assessment

Other Measures (see rubric on p. 162)

- Teacher observations

- STEM Research Notebook entries

- Participation in teams during investigations

INTERNET RESOURCES

Predicting earthquakes

- *www.usgs.gov/faqs/are-earthquake-probabilities-or-forecasts-same-prediction?qt-news_science_products=7#qt-news_science_products*

Earthquake magnitude scale

- *www.geo.mtu.edu/UPSeis/magnitude.html*

Earthquake background

- *https://earthquake.usgs.gov*

- *www.ready.gov/earthquakes*

- *https://disasters.nasa.gov/earthquakes*

- *www.kids.nationalgeographic.com/explore/science/earthquake/#earthquake-houses.jpg*

Tsunami detection technology

- *www.nasa.gov/feature/jpl/scientists-look-to-skies-to-improve-tsunami-detection*

Tsunami background
- *www.ready.gov/tsunamis*
- *www.natgeokids.com/uk/discover/geography/physical-geography/tsunamis*
- *http://tsunami.org/what-causes-a-tsunami*
- *http://news.bbc.co.uk/2/hi/asia-pacific/135797.stm*

Mount St. Helens volcano information
- *https://volcanoes.usgs.gov/volcanoes/st_helens*

Kīlauea volcano information
- *https://pubs.usgs.gov/fs/2013/3116*

Colli Albani volcano information
- *www.livescience.com/55397-extinct-rome-volcano-rumbles-to-life.html*

Volcano background
- *https://volcanoes.usgs.gov/index.html*
- *www.ready.gov/volcanoes*
- *www.spaceplace.nasa.gov/volcanoes2/en*
- *www.natgeokids.com/uk/discover/geography/physical-geography/volcano-facts*
- *www.geology.com/volcanoes/types-of-volcanic-eruptions*

Engineering professions
- *www.engineergirl.org/33/TryOnACareer*
- *www.nacme.org/types-of-engineering*
- *www.sciencekids.co.nz/sciencefacts/engineering/typesofengineeringjobs.html*

Earthquake-resistant structures
- *www.imaginationstationtoledo.org/educator/activities/can-you-build-an-earthquake-proof-building*
- *https://science.howstuffworks.com/engineering/structural/earthquake-resistant-buildings3.htm*

EDP resources
- *www.sciencebuddies.org/engineering-design-process/engineering-design-compare-scientific-method.shtml*

- *www.pbslearningmedia.org/resource/phy03.sci.engin.design.desprocess/what-is-the-design-process*

NOAA cost of natural disaster information
- *www.ncei.noaa.gov/news/calculating-cost-weather-and-climate-disasters*

Earthquake shake table instructions
- *www.pbskids.org/designsquad/build/seismic-shake-up*

Resources for student research on natural hazards

- Hurricane

  - *www.scied.ucar.edu/webweather/hurricanes*

  - *www.fema.gov/media-library-data/253331ac179d32652a4d0cbf7fb3e6eb/FEMA_FS_hurricanes_508_8-15-13.pdf*

- Tornado

  - *www.scied.ucar.edu/webweather/tornadoes*

  - *www.fema.gov/media-library-data/a4ec63524f9fd1fa5d72be63bd6b29cf/FEMA_FS_tornado_508_8-15-13.pdf*

- Earthquake

  - *https://earthquake.usgs.gov/learn/kids*

  - *www.fema.gov/media-library-data/ed51897eb6583e40ec1edb5f8fb85ee5/FEMA_FS_earthquake_508_081513.pdf*

- Tsunami

  - *www.ready.gov/kids/know-the-facts/tsunamis*

  - *www.fema.gov/media-library-data/26ea1ceee22e73558b01030a4a0d7c47/FEMA_FS_tsunami_508-8-15-13.pdf*

- Volcano

  - *www.ready.gov/kids/know-the-facts/volcano*

  - *www.fema.gov/media-library-data/a4402e44902b963c8de7ee4ad0586016/FEMA_FS_volcano_508-8-15-13.pdf*

Current earthquake data
- *https://earthquake.usgs.gov/earthquakes*

- *https://earthquake.usgs.gov/earthquakes/browse*

Map of U.S. regions
- *www2.census.gov/geo/pdfs/maps-data/maps/reference/us_regdiv.pdf*

A House Is a House for Me, read by author Mary Ann Hoberman
- *www.youtube.com/watch?v=s4cGB0-7R-0*

Lesson Plan 3: Our Natural Hazard Preparedness Plans

In this lesson, students work in teams to demonstrate their understanding of the impacts of natural hazards by creating preparedness plans describing how people and communities can prepare for natural hazards. Student teams also create public service announcements (PSAs) to communicate their preparedness plans.

ESSENTIAL QUESTIONS

- What impacts do natural hazards have on people and the environment?
- How can communities minimize damage related to the occurrence of natural hazards?
- How can people prepare for natural hazards in their communities?

ESTABLISHED GOALS AND OBJECTIVES

At the conclusion of this lesson, students will be able to do the following:

- Identify impacts of natural hazards on people and the environment
- Create a preparedness plan that can mitigate the impacts of a natural hazard on people and the environment
- Communicate their understanding of natural hazard preparedness through a PSA
- Understand that community characteristics can be expressed numerically and in text
- Organize numerical and textual information about their communities in an infographic
- Use technology to communicate about natural hazards
- Use technology tools to gather data about natural hazards

TIME REQUIRED

- 8 days (approximately 30 minutes each day; see Tables 3.9–3.10, pp. 39–40)

MATERIALS

Required Materials for Lesson 3

- STEM Research Notebooks

- Computer with internet access for viewing videos

- Smartphones, tablets, or a video recorder for recording teams' PSAs

- EDP graphic (optional handout; see p. 114)

- Book: *Flash, Crash, Rumble, and Roll,* by Franklyn M. Branley

- Chart paper

- Markers

- Map or globe

- Plain white paper

- Safety glasses with side shields or safety goggles (per student)

Additional Materials for Infographics (per team of 2–3 students unless otherwise indicated)

- Half sheet of poster board (14 × 22 inches)

- Set of markers

- 5 pieces of different-colored paper

- Glue

- Scissors (per student)

SAFETY NOTES

1. All students must wear sanitized safety glasses with side shields or safety goggles during all phases of this inquiry activity.

2. Use caution in working with sharps (scissors), as these can cut or puncture skin.

3. Keep glue away from eyes and skin.

4. Wash hands with soap and water after completing the cleanup phase of the activity.

CONTENT STANDARDS AND KEY VOCABULARY

Table 4.7 lists the content standards from the *NGSS, CCSS,* NAEYC, and the Framework for 21st Century Learning that this lesson addresses, and Table 4.8 (p. 102) presents the key vocabulary. Vocabulary terms are provided for both teacher and student use. Teachers may choose to introduce some or all of the terms to students.

Table 4.7. Content Standards Addressed in STEM Road Map Module
Lesson 3

NEXT GENERATION SCIENCE STANDARDS

PERFORMANCE EXPECTATIONS

- ETS1-1. Ask questions, make observations, and gather information about a situation people want to change to define a simple problem that can be solved through the development of a new or improved object or tool.

- ETS1-2. Develop a simple sketch, drawing, or physical model to illustrate how the shape of an object helps it function as needed to solve a given problem.

- ETS1-3. Analyze data from tests of two objects designed to solve the same problem to compare the strengths and weaknesses of how each performs.

SCIENCE AND ENGINEERING PRACTICES

Analyzing and Interpreting Data

Analyzing data in K–2 builds on prior experiences and progresses to collecting, recording, and sharing observations.

- Use observations (firsthand or from media) to describe patterns in the natural world in order to answer scientific questions.

Developing and Using Models

Modeling in K–2 builds on prior experiences and progresses to include using and developing models (i.e., diagram, drawing, physical replica, diorama, dramatization, storyboard) that represent concrete events or design solutions.

- Use a model to represent relationships in the natural world.

Planning and Carrying Out Investigations

Planning and carrying out investigations to answer questions or test solutions to problems in K–2 builds on prior experiences and progresses to simple investigations, based on fair tests, which provide data to support explanations or design solutions.

- Make observations (firsthand or from media) to collect data that can be used to make comparisons.

DISCIPLINARY CORE IDEAS

LS1.C. Organization for Matter and Energy Flow in Organisms

- All animals need food in order to live and grow. They obtain their food from plants or from other animals. Plants need water and light to live and grow.

ESS3.A. Natural Resources

- Living things need water, air, and resources from the land, and they live in places that have the things they need. Humans use natural resources for everything they do.

Continued

Table 4.7. (*continued*)

ESS2.D. Weather and Climate

- Weather is the combination of sunlight, wind, snow or rain, and temperature in a particular region at a particular time. People measure these conditions to describe and record the weather and to notice patterns over time.

CROSSCUTTING CONCEPTS

Patterns

- Patterns in the natural and human designed world can be observed, used to describe phenomena, and used as evidence.

Systems and System Models

- Systems in the natural and designed world have parts that work together.

Cause and Effect

- Events have causes that generate observable patterns.

COMMON CORE STATE STANDARDS FOR MATHEMATICS

MATHEMATICAL PRACTICES

- MP1. Make sense of problems and persevere in solving them.
- MP2. Reason abstractly and quantitatively.
- MP3. Construct viable arguments and critique the reasoning of others.
- MP4. Model with mathematics.
- MP5. Use appropriate tools strategically.
- MP6. Attend to precision.
- MP7. Look for and make use of structure.
- MP8. Look for and express regularity in repeated reasoning.

MATHEMATICAL CONTENT

- 2.NBT.A.1. Understand that the three digits of a three-digit number represent amounts of hundreds, tens, and ones; e.g., 706 equals 7 hundreds, 0 tens, and 6 ones.
- 2.NBT.A.2. Count within 1,000; skip-count by 5s, 10s, and 100s.
- 2.NBT.A.3. Read and write numbers to 1,000 using base-ten numerals, number names, and expanded form.

Continued

Table 4.7. (*continued*)

COMMON CORE STATE STANDARDS FOR ENGLISH LANGUAGE ARTS

READING STANDARDS

- RI.2.1. Ask and answer such questions as *who, what, where, when, why,* and *how* to demonstrate understanding of key details in a text.

- RI.2.3. Describe the connection between a series of historical events, scientific ideas or concepts, or steps in technical procedures in a text.

- RI.2.7. Explain how specific images (e.g., a diagram showing how a machine works) contribute to and clarify a text.

- RI.2.8. Describe how reasons support specific points the author makes in a text.

- RI.2.9. Compare and contrast the most important points presented by two texts on the same topic.

WRITING STANDARDS

- W.2.1. Write opinion pieces in which they introduce the topic or book they are writing about, state an opinion, supply reasons that support the opinion, use linking words (e.g., *because, and, also*) to connect opinion and reasons, and provide a concluding statement or section.

- W.2.2. Write informative/explanatory texts in which they introduce a topic, use facts and definitions to develop points, and provide a concluding statement or section.

- W.2.6. With guidance and support from adults, use a variety of digital tools to produce and publish writing, including in collaboration with peers.

- W.2.7. Participate in shared research and writing projects (e.g., read a number of books on a single topic to produce a report; record science observations).

- W.2.8. Recall information from experiences or gather information from provided sources to answer a question.

SPEAKING AND LISTENING STANDARDS

- SL.2.1. Participate in collaborative conversations with diverse partners about *grade 2 topics and texts* with peers and adults in small and larger groups.

- SL.2.2. Recount or describe key ideas or details from a text read aloud or information presented orally or through other media.

- SL.2.3. Ask and answer questions about what a speaker says in order to clarify comprehension, gather additional information, or deepen understanding of a topic or issue.

NATIONAL ASSOCIATION FOR THE EDUCATION OF YOUNG CHILDREN STANDARDS

- 2.G.02. Children are provided varied opportunities and materials to learn key content and principles of science.

- 2.G.03. Children are provided varied opportunities and materials that encourage them to use the five senses to observe, explore, and experiment with scientific phenomena.

Continued

Table 4.7. (*continued*)

- 2.G.04. Children are provided varied opportunities to use simple tools to observe objects and scientific phenomena.
- 2.G.05. Children are provided varied opportunities and materials to collect data and to represent and document their findings (e.g., through drawing or graphing).
- 2.G.06. Children are provided varied opportunities and materials that encourage them to think, question, and reason about observed and inferred phenomena.
- 2.G.07. Children are provided varied opportunities and materials that encourage them to discuss scientific concepts in everyday conversation.
- 2.G.08. Children are provided varied opportunities and materials that help them learn and use scientific terminology and vocabulary associated with the content areas.
- 2.H.02. All children have opportunities to access technology (e.g., tape recorders, microscopes, computers) that they can use.
- 2.H.03. Technology is used to extend learning within the classroom and integrate and enrich the curriculum.

FRAMEWORK FOR 21ST CENTURY LEARNING

- Interdisciplinary Themes; Learning and Innovation Skills; Information, Media, and Technology Skills; Life and Career Skills.

Table 4.8. Key Vocabulary for Lesson 3

Key Vocabulary	Definition
emergency	a potentially dangerous situation that is unexpected and requires people to take action to stay safe
infographic	a way of communicating information visually that uses graphic elements
issue	a problem or topic that is important to people
public service announcement (PSA)	a message about a social issue communicated to people through the media with the objective of providing information and persuading them to take some action; often related to health and safety
storyboard	a set of frames with information, ideas, or images that is used to plan a presentation in the order in which it will be presented

TEACHER BACKGROUND INFORMATION
Natural Hazard Preparedness

Students will learn about measures people and communities can take to prepare for natural hazards. The Federal Emergency Management Agency (FEMA) provides teacher and student resources that may be useful to you and your students throughout this lesson. These can be accessed at *www.ready.gov/kids*. You may also want to encourage students to create preparedness plans that pertain to natural hazards that are common in your area. The American Red Cross provides an interactive map that identifies the most common natural disasters for each region of the United States at *www.redcross.org/get-help/how-to-prepare-for-emergencies/common-natural-disasters-across-us.html*.

Public Service Announcements (PSAs)

Students will create PSAs as part of their final module challenge. Students should understand that PSAs are similar to the advertisements they usually see and hear on various media in that they use images and words to try to persuade their audience. They should also understand that PSAs are different from commercial advertisements because their goal is to persuade people to believe something or take action on issues typically related to health and safety, whereas commercial advertisements try to persuade people to buy things. PSAs can be televised, distributed on social media, announced on radio, or included in print publications. For this module's challenge, students will video record their own PSAs. Video PSAs use a variety of techniques to engage their viewers. These include using a "hook," such as a shocking image or humorous statement to get viewers' attention; communicating important information concisely, using language the audience can easily understand; and including a tagline that viewers will remember. PSAs on television can be especially impactful because they can reach broad audiences with powerful messages. Examples of classic, well-known PSAs include the Ad Council's Smokey Bear PSA (with the tagline "Only You Can Prevent Forest Fires") and the Partnership for a Drug-Free America's "This Is Your Brain on Drugs" PSA.

Creating PSAs provides students with an opportunity to actively engage with their learning and to use drama and storytelling to communicate information. Storyboarding, developing a series of images that depicts the flow of the PSA, can be a useful tool for students as they create PSAs and other video content. The following websites provide additional information about PSAs and storyboarding:

- *www.teachwriting.org/blog/2018/4/11/public-service-announcements-a-how-to-guide-for-teachers*

- *www.govtech.com/education/news/How-to-Create-the-Perfect-Public-Service-Announcement.html*

- *www.storyboardthat.com/articles/e/public-service-announcements*

COMMON MISCONCEPTIONS

Students will have various types of prior knowledge about the concepts introduced in this lesson. Table 4.9 outlines some common misconceptions students may have concerning these concepts. Because of the breadth of students' experiences, it is not possible to anticipate every misconception that students may bring as they approach this lesson. Incorrect or inaccurate prior understanding of concepts can influence student learning in the future, however, so it is important to be alert to misconceptions such as those presented in the table.

Table 4.9. Common Misconceptions About the Concepts in Lesson 3

Topic	Student Misconception	Explanation
Natural hazards	Nothing can be done to prepare for a natural hazard, since they are caused by nature and occurrences are unpredictable.	While it is not possible to stop natural hazards from occurring, people and communities can take measures to be prepared, including building structures with design features that have been created to withstand natural hazards and making and communicating plans for what people should do in the event that a natural hazard strikes their community.
	The images we see in the media accurately show the impacts of natural hazards.	The media tend to focus on sensational images and information that will capture consumers' attention. This means that the public's perception of some natural hazard events and damage associated with them might be skewed toward the extreme impacts and not provide an overall picture of community impact.

PREPARATION FOR LESSON 3

Review the Teacher Background Information section (p. 103), assemble the materials for the lesson, and preview the video recommended in the Learning Components section below. Identify a current or well-known public service announcement video to show the class in the Introductory Activity/Engagement.

In this lesson's mathematics and social studies connections, students will create descriptions of their town or city using numerical data, so you should access some basic descriptive information such as population and the number of homes, businesses, schools, and public parks. Prepare this information in narrative form. You should also include some nonquantifiable information in the narrative. Here's an example: "The town of Smithville has 40,000 residents. It has been named a Tree City USA because of the 50 beautiful oak trees in the town. Last year, about 5,000 people came from all over the country to see the leaves in the autumn. The town has four grocery stores, four hotels, and three public parks. The distance from the west side of town to the east side is 4 miles, and the distance from the south side of town to the north side is 2 miles." Students will prepare simple infographics with this information to complement their natural hazard preparedness plans. You should have images of some simple infographics on hand to share with students. These can be identified by conducting an internet image search using terms such as "examples of infographics" and "infographics for elementary students." Identify examples that include text and graphical representations of information.

Students will conduct additional research on natural hazard preparedness (see STEM Research Notebook Entry #25) to create their preparedness plans. There are numerous online and printed resources that may be useful, including several of the websites provided in the Teacher Background Information section in Lesson 1 (see pp. 51–56) and the websites and books listed below. You should either bookmark websites on a computer with internet access for each team or print out the online information. Student resources about natural hazard preparation include the following:

- *www.ready.gov/kids/know-the-facts*
- *www.fema.gov/media-library/assets/documents/34288*
- *Violent Weather: Thunderstorms, Tornadoes, and Hurricanes,* by Andrew Collins, page 33
- *Flash, Crash, Rumble, and Roll,* by Franklyn M. Branley, pages 22–25
- *Hurricanes!* by Gail Gibbons, pages 28–31
- *Eye of the Storm: A Book About Hurricanes,* by Rick Thomas, page 22
- *Tornado Alert,* by Franklyn M. Branley, pages 24–32
- *Earthquakes,* by Ellen Prager, pages 26–27
- *Earthquakes,* by Franklyn M. Branley, pages 28–30

Students will use the steps of the EDP to prepare this plan and the public service announcement. The STEM Research Notebook pages provided in this lesson will facilitate students' use of the EDP, and you should prepare by posting the EDP graphic provided on page 114 in the classroom or making copies of the graphic for each team. Students will plan their PSAs by creating storyboard frames. Each student will create two or more frames using STEM Research Notebook Entry #27, which includes a single

storyboard frame template. The class will decide what types of information should be provided in the PSAs (see Activity/Exploration), and this, along with the number of students on each team, will determine how many frames each student will prepare. You should make one copy for each storyboard frame each student will be responsible for.

You should also prepare for video recording of student PSAs, either by providing teams with individual devices or by setting up a station with a video camera on a tripod. You should also identify where students will video their PSAs (e.g., outdoors or indoors against a plain-colored backdrop) and plan accordingly. As a culminating activity for the module, students will view all PSAs. Prepare to have equipment on hand to show the PSAs. Make appropriate preparations if you choose to share the PSAs with other classes at the school or invite parents to participate in the presentation day.

LEARNING COMPONENTS
Introductory Activity/Engagement

Connection to the Challenge: Begin each day of this lesson by directing students' attention to the module challenge, the Natural Hazard Preparedness Challenge:

> *Your town's leaders want to be sure that people will be safe in case a natural hazard should strike your town. They have asked your class to create preparedness plans for natural hazards that will keep people safe during these events. You and your team are challenged to create a plan for the community for one natural hazard and an advertisement that lets people know about your plan.*

Hold a brief class discussion of how students' learning in the previous days' lessons contributed to their ability to complete the challenge. You may wish to create a class list of key ideas on chart paper.

Science Class: Hold a class discussion of emergency preparedness by asking questions such as the following:

- How can natural hazards affect people?

- How can natural hazards affect communities?

- How can natural hazards affect the environment?

- How can communities minimize the impact of the damage caused by natural hazards on people? On communities? On the environment?

- What can people do to prepare for natural hazards in their community?

- What can communities do to prepare for natural hazards?

ELA Connection: Ask students for their ideas about what a PSA is. Show students the PSA you identified (see p. 104). Ask students what they think the purpose of this advertisement is. Next, ask students how this is the same as and different from the advertisements they typically see on television, recording students' responses on a class T-chart. Introduce the idea that a PSA is a special type of advertisement that is meant to educate people about an issue that is important for their health or safety. Tell students that as part of their challenge, they will create PSAs of the natural hazard preparedness plans they create.

Have students view an informational emergency preparedness video created by elementary-level students, such as "Ready, Set, Prepare" at *www.youtube.com/ watch?v=fp1DyJsuQwU*. Ask students to share what they know and what they learned from the video about natural hazard preparation. Record students' responses on a KLEWS chart.

STEM Research Notebook Entry #23

Have students document what they learned about natural hazard preparedness and public service announcements in their STEM Research Notebooks after viewing the video, using both words and pictures.

Mathematics and Social Studies Connections: Remind students that in the last lesson, they used bar graphs to display information about earthquakes. Ask students to share what they know about the benefits of using bar graphs to display information. Next, hand out the narrative that you prepared with information about your town or city (see pp. 104–105). Read aloud the narrative as a class. After reading, ask students what they learned about their town or city. Emphasize to students that this information provides a description of their town or city using numbers as well as words. Create a table on chart paper with three columns, labeling them "Numbers," "Words," and "Words and Numbers." Ask students to identify the information described only in numbers, the information described only in words, and the information described in both numbers and words, and enter this information into the table.

Then, point to the "Numbers" column and ask students if they could easily display this information as a bar graph and how they could do this. Emphasize to students that bar graphs are good for displaying information with two features (e.g., earthquake location and earthquake strength) but will not work well to display more complex information or pieces of information that do not use the same units (for example, number of people in a town and distance from the north end to the south end). Point to the column labeled "Words," and ask students if this information could be presented on a bar graph.

Next, have students work in their teams to brainstorm ideas about ways that the information about their town provided in the narrative could be displayed visually. Tell

students that they will use this information about their towns in their emergency preparedness plans to help explain why it is important to protect people and the environment in their community. Have teams share their ideas with the class. Introduce the idea that an infographic is a way to visually display several pieces of information about a topic using both words and numbers. Show the examples of infographics you prepared, and ask students how each piece of information was displayed and how the infographics are organized (e.g., in sections). As a class, brainstorm ideas about how each piece of information in the narrative could be displayed on an infographic, creating a class list of students' ideas. Have students work in their teams to create infographics on poster board that display the information provided in the narrative.

Activity/Exploration

Science Class and Mathematics, ELA, and Social Studies Connections: Ask students to share their ideas about what the word *emergency* means. Create a class definition that reflects that an emergency is unexpected, is possibly dangerous, and requires quick action. Ask students how they would know if there is an emergency situation, creating a class list (e.g., alarms, television announcements, radio announcements, messages on phones, tornado or other emergency sirens). Ask students to share their ideas about why they are being told about an emergency, and guide students to understand that they are provided with this information so that they can take action to protect themselves and others.

Ask students to name some potentially dangerous events associated with natural hazards (e.g., high winds, flooding). Point out to students that some of these require extreme action, such as evacuating or placing sandbags around homes to keep water out, but that there are other hazards, such as lightning, that require action that may not be as extreme. Ask students to name ways that they can stay safe in the case of a thunderstorm, recording student ideas on a KLEWS chart. Have the class explore thunderstorms and emergency preparation by participating in an interactive read-aloud of *Flash, Crash, Rumble, and Roll,* by Franklyn M. Branley. Document student learning on the KLEWS chart.

STEM Research Notebook Entry #24

Have students document what they learned about thunderstorms and what they can do to prepare for and stay safe in storms in their STEM Research Notebooks, using both words and pictures.

Ask students to name the steps of the EDP, and tell them that they will use this process to prepare their plans and PSAs. Point to the Define step on the EDP graphic, and ask students to define what problem they are trying to solve (how to help people be prepared for a natural hazard). Now point to the Learn step on the EDP graphic, and ask students

what they do during the Learn step (research, learn about the problem, and brainstorm ideas). Tell students that they already started the Learn step by researching their team's natural hazard in Lesson 2 and preparing an infographic about their town. Students will use the posters they prepared in Lesson 2 to provide basic information about their team's natural hazard and some information about how to stay safe during that hazard. Now, tell students that they will do some additional research on how to prepare for the natural hazard using the graphic organizer provided in STEM Research Notebook Entry #25.

STEM Research Notebook Entry #25

Have students document their team's natural hazard and the problem they are trying to solve. They should complete the graphic organizer with information about how to prepare for the natural hazard their teams were assigned, using the information from the Hazard Sleuths poster from Lesson 2 as well as the additional resources provided on page 95 to begin devising a natural hazard preparedness plan.

After teams finish gathering their background information, tell students that for the Plan step of the EDP, teams will each create a storyboard that shows the important information for the public service announcement and the order in which it should be presented. Explain to students that a storyboard is a tool used to plan for video productions to help their creators make sure they reach their goal. Tell students that each storyboard frame will present an important piece of information and describe how it will be presented. As a class, brainstorm ideas about what information should be provided in the PSAs, creating a class list. At a minimum, the PSAs should include the following:

- The name of the hazard

- Information about the cause of the hazard

- Information about your town or city and how this natural hazard could affect it

- How people are informed about the hazard

- What they can do in advance to prepare for the hazard

- What they can do during the natural hazard event to stay safe

- How they can protect property during the natural hazard

Next, ask students for their ideas about how they could present this information (e.g., using their infographics, using their team posters, showing photos, acting out things people could do to prepare, creating a song that people will remember). Create a class list of student ideas, and have the class decide on a final list of types of information that should be included in the PSAs.

STEM Research Notebook Entry #26

Have students enter these types of information in the first column of the table provided in STEM Research Notebook Entry #26. Then, have teams decide which member will present what type of information and how it should be presented, entering these in the second and third columns of the table.

Each student on the team should create two or more storyboard frames (depending on the number of students on the team and the number of items the class decided should be included). Students will be responsible for sharing the information in their storyboard frames in the public service announcement.

Explanation

Science Class and Mathematics, ELA, and Social Studies Connections: For the Try step, students will create their storyboards. Students should use one copy of STEM Research Notebook Entry #27 for each frame they will prepare. Hand out plain white paper to each student (one sheet for each storyboard frame he or she is responsible for). Remind students to use the table they created in STEM Research Notebook Entry #26 to identify which team members are responsible for what type of information and how this information should be presented. Tell students they should limit their PSAs to seven minutes or less, so students should spend less than one minute presenting each storyboard frame.

STEM Research Notebook Entry #27

Have students each complete one storyboard frame for each type of information he or she is responsible for in the team's PSA.

Then, have teams practice presenting their PSAs as outlined on the storyboard frames. Remind students that their PSAs should be seven minutes or less in length. Remind students that the Test step of the EDP is important for making improvements to their presentations. Tell them that before they record their PSAs, they will present them to other students to receive feedback. Pair teams, and have one team present its PSA while the members of the second team in the pair provide feedback using STEM Research Notebook Entry #28. Then, have teams switch roles so that each team receives feedback from members of another team.

STEM Research Notebook Entry #28

Have each student provide feedback about another team's PSA in his or her STEM Research Notebook.

For the Decide step of the EDP, have teams review the feedback they received from other students and decide what changes they should make based on that feedback.

STEM Research Notebook Entry #29

Have students document in their STEM Research Notebooks the changes their team decided to make based on the feedback the team received on its PSA presentation.

Elaboration/Application of Knowledge

Science Class and Mathematics, ELA, and Social Studies Connections: Have a presentation day when the class views all teams' PSAs. You may also consider inviting parents to attend or having students share their PSAs with other classes in the school. After viewing each PSA, hold a class discussion about the strengths and weaknesses of each team's natural hazard preparedness plan. After all presentations have been made, have teams meet to discuss ways that they could improve their plans based on the class discussion.

STEM Research Notebook Entry #30

Have students document in their STEM Research Notebooks their teams' ideas about how their natural hazard preparedness plans could be improved.

Next, review the collaboration rules the class formulated in Lesson 2. Hold a class discussion about how teams collaborated during the Natural Hazard Preparedness Challenge, asking students to reflect on what went well and what was challenging.

STEM Research Notebook Entry #31

Have students create a STEM Research Notebook entry with their individual reflections about how their team collaborated during the Natural Hazard Preparedness Challenge.

Assess student learning in this lesson by asking students to do the following:

- Identify four types of natural hazards and the cause of each
- Identify five ways natural hazards can have an impact on people
- Identify three ways natural hazards can have an impact on the environment
- Describe five ways people can prepare for natural hazards

Evaluation/Assessment

Students may be assessed on the following performance tasks and other measures listed.

Performance Tasks

- Community infographics
- Our Natural Hazard Preparedness Plans public service announcements
- Lesson assessment

Other Measures (see rubric on p. 162)

- Teacher observations
- STEM Research Notebook entries
- Participation in teams

INTERNET RESOURCES

Red Cross information about natural disasters by region

- *www.redcross.org/get-help/how-to-prepare-for-emergencies/common-natural-disasters-across-us.html*

FEMA natural hazard preparedness resources

- *www.ready.gov/kids*

PSAs and storyboarding

- *www.teachwriting.org/blog/2018/4/11/public-service-announcements-a-how-to-guide-for-teachers*

- *www.govtech.com/education/news/How-to-Create-the-Perfect-Public-Service-Announcement.html*

- *www.storyboardthat.com/articles/e/public-service-announcements*

Student resources for natural hazard preparedness

- *www.ready.gov/kids/know-the-facts*

- *www.fema.gov/media-library/assets/documents/34288*

Sample student-created PSA

- *www.youtube.com/watch?v=fp1DyJsuQwU*

SUGGESTED BOOKS

- *Danger! Earthquakes,* by Seymour Simon (SeaStar Books, 2002)

- *Earthquakes,* by Franklyn M. Branley (HarperCollins, 2015)

- *Earthquakes,* by Paul P. Sipiera (Children's Press, 1999)

- *Eye of the Storm: A Book About Hurricanes,* by Rick Thomas (Picture Window Books, 2005)

- *Heat Waves and Droughts,* by Liza N. Burby (PowerKids Press, 1999)

- *How Much Is a Million?* by David M. Schwartz (HarperCollins, 2004)

- *Hurricanes!* by Gail Gibbons (Holiday House, 2009)

- *Hurricanes,* by Seymour Simon (HarperCollins, 2007)

- *If You Made a Million,* by David M. Schwartz (HarperCollins, 1994)

- *On Beyond a Million: An Amazing Math Journey,* by David M. Schwartz (Doubleday Books for Young Readers, 1999)

- *Tornado Alert,* by Franklyn M. Branley (HarperCollins, 1988)

- *Tornadoes!* by Gail Gibbons (Holiday House, 2009)

- *Tornadoes,* by Seymour Simon (HarperCollins, 2001)

- *Twisters and Other Terrible Storms,* by Will Osborne and Mary Pope Osborne (Random House, 2003)

REFERENCE

Koehler, C., M. A. Bloom, and A. R. Milner. 2015. The STEM Road Map for grades K–2. In *STEM Road Map: A framework for integrated STEM education,* ed. C. C. Johnson, E. E. Peters-Burton, and T. J. Moore, 41–67. New York: Routledge. *www.routledge.com/products/9781138804234.*

ENGINEERING DESIGN PROCESS

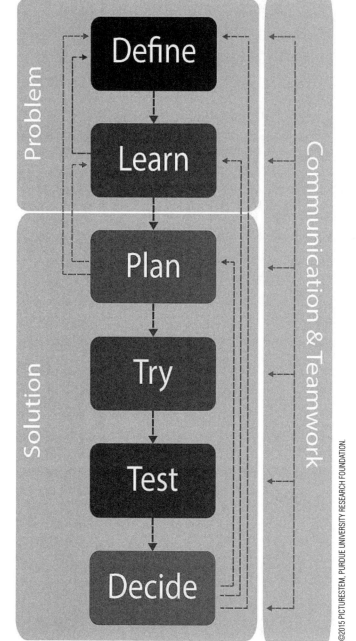

TRANSFORMING LEARNING WITH NATURAL HAZARDS AND THE *STEM ROAD MAP CURRICULUM SERIES*

Carla C. Johnson

This chapter serves as a conclusion to the Natural Hazards integrated STEM curriculum module, but it is just the beginning of the transformation of your classroom that is possible through use of the *STEM Road Map Curriculum Series*. In this book, many key resources have been provided to make learning meaningful for your students through integration of science, technology, engineering, and mathematics, as well as social studies and English language arts, into powerful problem- and project-based instruction. First, the Natural Hazards curriculum is grounded in the latest theory of learning for students in grade 2 specifically. Second, as your students work through this module, they engage in using the engineering design process (EDP) and build prototypes like engineers and STEM professionals in the real world. Third, students acquire important knowledge and skills grounded in national academic standards in mathematics, English language arts, science, and 21st century skills that will enable their learning to be deeper, retained longer, and applied throughout, illustrating the critical connections within and across disciplines. Finally, authentic formative assessments, including strategies for differentiation and addressing misconceptions, are embedded within the curriculum activities.

The Natural Hazards curriculum in the Cause and Effect STEM Road Map theme can be used in single-content classrooms (e.g., science) where there is only one teacher or expanded to include multiple teachers and content areas across classrooms. Through the exploration of the Natural Hazard Preparedness Challenge, students engage in a real-world STEM problem on the first day of instruction and gather necessary knowledge and skills along the way in the context of solving the problem.

The other topics in the *STEM Road Map Curriculum Series* are designed in a similar manner, and NSTA Press has additional volumes in this series for this and other grade levels and plans to publish more. The volumes covering Innovation and Progress have been published and are as follows:

- *Amusement Park of the Future, Grade 6*

- *Construction Materials, Grade 11*

- *Harnessing Solar Energy, Grade 4*

- *Transportation in the Future, Grade 3*

- *Wind Energy, Grade 5*

The volumes covering the Represented World have also been published and are as follows:

- *Car Crashes, Grade 12*

- *Improving Bridge Design, Grade 8*

- *Investigating Environmental Changes, Grade 2*

- *Packaging Design, Grade 6*

- *Patterns and the Plant World, Grade 1*

- *Radioactivity, Grade 11*

- *Rainwater Analysis, Grade 5*

- *Swing Set Makeover, Grade 3*

The tentative list of other books includes the following themes and subjects:

- Cause and Effect
  - Earth on the move
  - Healthy living
  - Human impacts on our climate
  - Influence of waves
  - Physics in motion
- Sustainable Systems
  - Composting: Reduce, reuse, recycle
  - Creating global bonds

- Hydropower efficiency

- System interactions

• Optimizing the Human Experience

- Genetically modified organisms

- Mineral resources

- Rebuilding the natural environment

- Water conservation: Think global, act local

If you are interested in professional development opportunities focused on the STEM Road Map specifically or integrated STEM or STEM programs and schools overall, contact the lead editor of this project, Dr. Carla C. Johnson (*carlacjohnson@ncsu.edu*), associate dean and professor of science education and executive director of the William and Ida Friday Institute at North Carolina State University. Someone from the team will be in touch to design a program that will meet your individual, school, or district needs.

APPENDIX A

STEM RESEARCH NOTEBOOK TEMPLATES

MY STEM RESEARCH NOTEBOOK

NATURAL HAZARDS

Name:

_____

- -

_____

Name: _____   Date: _____

STEM RESEARCH NOTEBOOK ENTRY #1 (LESSON PLAN 1)

NATURAL HAZARD

Draw and label a natural hazard or memorable weather event that you either have experienced or know about.

Write a Description	Draw and Label

Name: _____   Date: _____

STEM RESEARCH NOTEBOOK ENTRY #2, PAGE 1 (LESSON PLAN 1)

CURRENT NATURAL DISASTER

1. Identify the current natural disaster that the class is following. (Does it have a "name"?)

   _____

   _____

   _____

   _____

   _____

2. When did this natural disaster occur?

   _____

   _____

   _____

   _____

   _____

3. Does this kind of disaster happen most often at a certain time of the year? If so, when?

   _____

   _____

   _____

   _____

   _____

4. Where in the world is this natural disaster happening?

   _____

   _____

   _____

   _____

   _____

Name: _____   Date: _____

STEM RESEARCH NOTEBOOK #2, PAGE 2 (LESSON PLAN 1)

CURRENT NATURAL DISASTER

5. Where in the world does this disaster most commonly occur?

_____

_____

_____

_____

6. How does this natural disaster happen?

_____

_____

_____

_____

_____

_____

_____

_____

Name: _____   Date: _____

STEM RESEARCH NOTEBOOK ENTRY #3 (LESSON PLAN 1)

NATURAL HAZARDS AND THEIR CAUSES

Weather	Movement Within Earth

Name: _____   Date: _____

STEM RESEARCH NOTEBOOK ENTRY #4, PAGE 1 (LESSON PLAN 1)

I learned ...

Thunderstorms are caused by:

_____

- -

_____

_____

- -

_____

Tornadoes are caused by:

_____

- -

_____

_____

- -

_____

Hurricanes are caused by:

_____

- -

_____

_____

- -

_____

Name: _____   Date: _____

STEM RESEARCH NOTEBOOK ENTRY #4, PAGE 2 (LESSON PLAN 1)

Ways Thunderstorms, Tornadoes, and Hurricanes Are the Same	Ways Thunderstorms, Tornadoes, and Hurricanes Are Different

Name: _____

Date: _____

STEM RESEARCH NOTEBOOK ENTRY #5 (LESSON PLAN 1)

VOCABULARY WORDS

Key Vocabulary Word	Definition	Illustration

Name: _____   Date: _____

VORTEX BOTTLE

- First, circle or write in your predictions.
- After you create your vortex bottle, circle or write in your observations.
- Then, sketch and label your observations.

What will happen to the debris in the vortex bottle ...	Predictions	Observations
While the water is spinning in a vortex?	Nothing It will float to the top It will float to the bottom Other: _____ _____	Nothing happened It floated to the top It floated to the bottom Other: _____ _____
After the water is no longer spinning in a vortex?	Nothing It will float to the top It will float to the bottom Other: _____ _____	Nothing happened It floated to the top It floated to the bottom Other: _____ _____

Name: _____   Date: _____

VORTEX BOTTLE
—SKETCHES—

While the water is spinning in a vortex	After the water is no longer spinning in a vortex

Name: _____   Date: _____

STEM RESEARCH NOTEBOOK ENTRY #8 (LESSON PLAN 1)

I learned ...

Name: _____

Date: _____

STEM RESEARCH NOTEBOOK ENTRY #9 (LESSON PLAN 1)

EXPLANATIONS

What happened to the debris in the vortex bottle …	Cause	Effect
While the water was spinning in a vortex?		
After the vortex stopped spinning and the water was still?		

Name: _____   Date: _____

STEM RESEARCH NOTEBOOK ENTRY #10 (LESSON PLAN 1)

I learned ...

_____

_____

_____

_____

_____

_____

_____

_____

Name: _____   Date: _____

STEM RESEARCH NOTEBOOK ENTRY #11 (LESSON PLAN 1)

MY WEATHER TALL TALE

I woke up one Saturday morning and heard wind outside. I looked out my window and saw …

Basic Parts of a Story	Parts of My Story

Name: _____   Date: _____

STEM RESEARCH NOTEBOOK ENTRY #12 (LESSON PLAN 2)

I learned ...

NATIONAL SCIENCE TEACHING ASSOCIATION

Name: _____

Date: _____

STEM RESEARCH NOTEBOOK ENTRY #13 (LESSON PLAN 2)

ANIMAL HOMES

Kind of Animal and Its Home	Impact on Animal Homes From Natural Hazards		
	Tornado	Flood	Earthquake

Name: _____   Date: _____

STEM RESEARCH NOTEBOOK ENTRY #14 (LESSON PLAN 2)

FINANCIAL IMPACT OF THE CURRENT NATURAL DISASTER

Type of Impact	Cost

Name: _____   Date: _____

STEM RESEARCH NOTEBOOK ENTRY #15 (LESSON PLAN 2)

OUR COLLABORATION CONTRACT

I agree to be a good team member by following these rules:

1. _____

   _____

   _____

2. _____

   _____

   _____

3. _____

   _____

   _____

4. _____

   _____

   _____

Signed: _____

Name: _____   Date: _____

STEM RESEARCH NOTEBOOK ENTRY #16 (LESSON PLAN 2)

EARTHQUAKE SHAKE

DEFINE: What is the problem you are trying to solve?

_____

- -

_____

_____

- -

_____

LEARN: What information do you need to solve this problem?

_____

- -

_____

_____

- -

_____

PREDICT: Write your predictions.

Question	Predictions
What materials will make the most earthquake-proof building?	

Name: _____   Date: _____

STEM RESEARCH NOTEBOOK ENTRY #17, PAGE 1 (LESSON PLAN 2)

PLAN: Look at the materials available, and choose the materials you think will create the most earthquake-proof structure. Next, plan a design for your structure. Draw a picture of it here, and label the materials you chose.

[ ]

TRY: Build your structure using your plan as a guide. Does it look like the drawing in your plan? Make a drawing of your structure before the earthquake.

[ ]

Name: _____   Date: _____

STEM RESEARCH NOTEBOOK ENTRY #17, PAGE 2 (LESSON PLAN 2)

TEST (Observe): What happens when the "earthquake" occurs? How long did your structure stand? Draw a picture of what happened, showing what your structure looked like after the earthquake.

Name: _____   Date: _____

STEM RESEARCH NOTEBOOK ENTRY #18 (LESSON PLAN 2)

DECIDE (Explain): Can you improve on your structure's design? How?

_____

- -

_____

_____

- -

_____

Draw a picture of your best design.

Why do you think this design is the best one?

_____

- -

_____

_____

_____

- -

_____

_____

- -

_____

Name: _____   Date: _____

STEM RESEARCH NOTEBOOK ENTRY #19 (LESSON PLAN 2)

BAR GRAPH OF CURRENT EARTHQUAKES BY CONTINENT

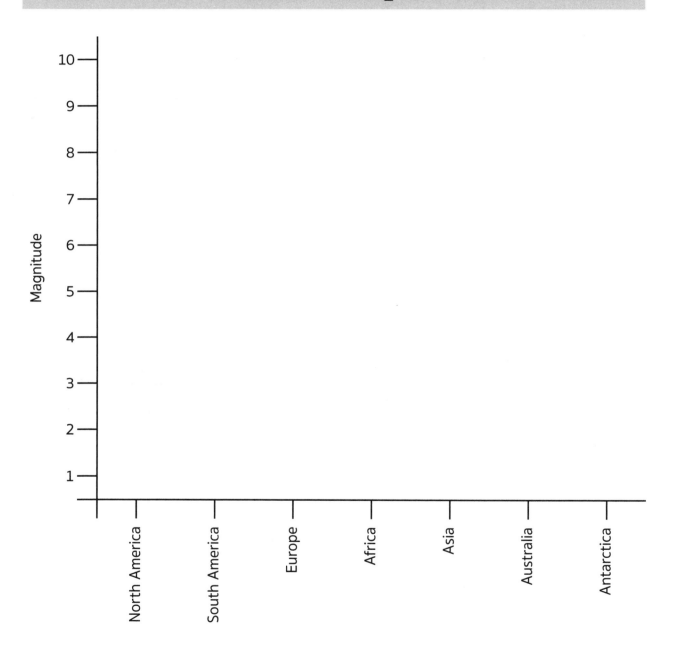

Name: _____   Date: _____

HAZARD SLEUTHS
—TEAM RESEARCH ON NATURAL DISASTER—

1. Identify the natural hazard that your group is researching:

   _____

   _____

   _____

   _____

2. Does this hazard happen most often at a certain time of the year? If so, when? Why?

   _____

   _____

   _____

   _____

   _____

   _____

   _____

   _____

Name: _____   Date: _____

STEM RESEARCH NOTEBOOK ENTRY #20, PAGE 2 (LESSON PLAN 2)

HAZARD SLEUTHS

3. Where in the world does this hazard most commonly occur?

_____

- -

_____

_____

- -

_____

4. What causes this natural hazard to happen?

_____

- -

_____

_____

- -

_____

_____

- -

_____

_____

- -

_____

Name: _____   Date: _____

STEM RESEARCH NOTEBOOK ENTRY #20, PAGE 3 (LESSON PLAN 2)

HAZARD SLEUTHS

5. How can this natural hazard affect a community?

_____

- -

_____

_____

- -

_____

_____

- -

_____

Draw and label a picture of the impacts this hazard can have on a community.

Name: _____   Date: _____

STEM RESEARCH NOTEBOOK ENTRY #20, PAGE 4 (LESSON PLAN 2)

HAZARD SLEUTHS

6. How can people stay safe if this hazard strikes their community?

_____

_____

_____

_____

_____

_____

_____

_____

_____

_____

_____

_____

_____

_____

_____

_____

Name: _____   Date: _____

BAR GRAPH OF CURRENT EARTHQUAKES BY U.S. REGION

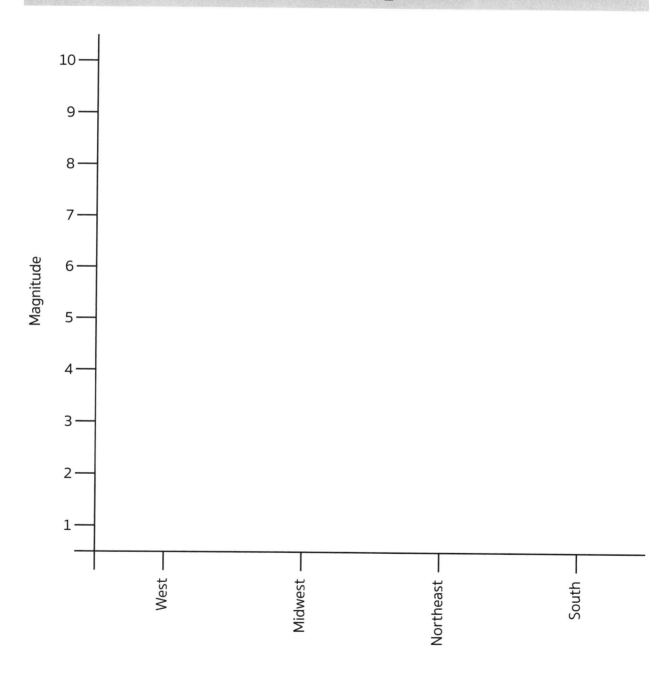

Name: _____   Date: _____

STEM RESEARCH NOTEBOOK ENTRY #22 (LESSON PLAN 2)

"ANIMALS IN A NATURAL HAZARD" STORY

Basic Parts of a Story	Parts of My Story

Name: _____   Date: _____

STEM RESEARCH NOTEBOOK ENTRY #23 (LESSON PLAN 3)

I learned …

Name: _____   Date: _____

STEM RESEARCH NOTEBOOK ENTRY #24 (LESSON PLAN 3)

I learned ...

Name: _____   Date: _____

STEM RESEARCH NOTEBOOK ENTRY #25, PAGE 1 (LESSON PLAN 3)

PREPARING FOR A NATURAL HAZARD
(DEFINE AND LEARN)

My team's natural hazard is: _____

The problem we are trying to solve is: _____

_____

_____

HOW MUCH TIME WILL WE HAVE TO PREPARE?

Can we watch a weather forecast to know that this natural hazard is possible?

How will we know that this natural hazard is about to occur?

We will have this amount of time to prepare for this natural hazard (circle one):

Days   Hours   Minutes

Name: _____   Date: _____

STEM RESEARCH NOTEBOOK ENTRY #25, PAGE 2 (LESSON PLAN 3)

PREPARING FOR A NATURAL HAZARD —PROTECTING PEOPLE—

Is there a special place in their homes people should go if this hazard occurs? Where?

_____

_____

What should be in people's emergency kits?

_____

_____

_____

_____

_____

Should families leave their homes if we know this natural hazard is going to occur?

_____

Where should families go if they leave their homes?

_____

_____

Name: _____   Date: _____

PREPARING FOR A NATURAL HAZARD
—PROTECTING PROPERTY—

What are three ways people can protect their homes from this natural hazard?

1. _____

_____

_____

_____

2. _____

_____

_____

_____

3. _____

_____

_____

_____

Name: _____ Date: _____

STEM RESEARCH NOTEBOOK ENTRY #26 (LESSON PLAN 3)

PUBLIC SERVICE ANNOUNCEMENT (PLAN)

We will include the following information in our PSA:

Type of Information for PSA	Team Member Responsible	Ideas for How This Can Be Presented

Type of Information for PSA	Team Member Responsible	Ideas for How This Can Be Presented

Name: _____   Date: _____

STEM RESEARCH NOTEBOOK ENTRY #27 (LESSON PLAN 3)

STORYBOARD FRAME
(TRY)

Type of information for this storyboard frame: _____

I will introduce the information I am providing by saying:

This is the information I will provide:

I will present the information in this way:

Name: _____   Date: _____

PSA FEEDBACK
(TEST)

This feedback is for the following team (list team members' names):

_____

_____

The things I like best about this team's PSA are:

_____

_____

_____

_____

This PSA could be improved by:

_____

_____

_____

_____

_____

_____

Name: _____   Date: _____

STEM RESEARCH NOTEBOOK ENTRY #29 (LESSON PLAN 3)

IMPROVEMENTS TO OUR PSA
(DECIDE)

After reading the feedback for our team's presentation, we decided to improve our PSA by:

_____

_____

_____

_____

_____

_____

_____

_____

_____

_____

_____

_____

Name: _____   Date: _____

RESPONSE TO CLASS FEEDBACK ON PSA AFTER PRESENTATION

After presenting our team's natural hazard preparedness plan and hearing the class's feedback, we think we could improve our PSA by:

_____

_____

_____

_____

_____

_____

_____

_____

_____

_____

_____

_____

Name: _____   Date: _____

STEM RESEARCH NOTEBOOK ENTRY #31 (LESSON PLAN 3)

REFLECTION ON COLLABORATION

Describe how you think your team collaborated during the Natural Hazard Preparedness Challenge.

I think our team's collaboration was (circle one):

Excellent   Pretty good   Just OK   Not very good

I think our team followed the class collaboration rules (circle one):

Always   Most of the time   Sometimes   Almost never

I think our team worked best together when we:

_____

_____

_____

_____

Our team struggled to collaborate when we:

_____

_____

_____

_____

APPENDIX B

ASSESSMENT RUBRICS

Name: _____

Observation, STEM Research Notebook, and Participation Rubric

Categories (Components)	Missing or Unrelated (0 points)	Beginning (1 point)	Developing (2 points)	Meets Expectations (3 points)	Exceeds Expectation (4 points)	Score
OBSERVATION OF LISTENING AND DISCUSSION SKILLS	Component is missing or unrelated.	Does not listen to others, and shows little respect for alternative viewpoints.	Occasionally listens to others but often speaks out of turn.	Listens to others, only occasionally speaks out of turn, and generally accepts other points of view.	Listens carefully to others, waits for turn to speak, and respects alternative viewpoints.	
STEM RESEARCH NOTEBOOK	Component is missing or unrelated.	Indicates little understanding of the concepts being taught.	Recalls and is able to explain basic facts and concepts.	Demonstrates ability to apply concepts, using information in new situations.	Demonstrates a deep understanding of concepts by drawing relationships between ideas and using information to generate new ideas.	
PARTICIPATION	Component is missing.	Does not volunteer. When responding to teacher prompts, comments are sometimes not relevant to the discussion.	Responds to teacher prompts during classroom discussions but does not volunteer.	Willingly participates in classroom discussions and offers relevant comments.	Contributes insightful comments and poses thoughtful questions.	

TOTAL SCORE: _____

COMMENTS:

Creative Writing Rubric

Name: _____

Categories (Components)	Missing or Unrelated (0 points)	Beginning (1 point)	Developing (2 points)	Meets Expectations (3 points)	Exceeds Expectations (4 points)	Score
STORY STRUCTURE	Component is missing or unrelated.	The story has little structure, and two out of the three components (beginning, middle, or end) are missing or are not related to the story's plot.	The story is missing one component of structure (beginning, middle, or end), or the components are not related to the story's plot.	The story has a beginning, middle, and end that relate to the story's plot.	The story has a clearly presented beginning, middle, and end that focus on the story's plot.	
STORY ELEMENTS	Component is missing or unrelated.	No plot is present. Characters or setting are not used or are difficult to identify. No clear solution is presented.	The plot may be confusing and the setting vague. Characters may not be introduced or identified, or there may be no solution.	The plot is understandable and presented in a specific setting. Characters are identified, and a solution is presented.	The plot is clearly and creatively presented in a specific and well-described setting. Characters are well developed, and a solution develops from the interaction of the story elements.	

Continued

Creative Writing Rubric (*continued*)

Categories (*Components*)	Missing or Unrelated (0 points)	Beginning (1 point)	Developing (2 points)	Meets Expectations (3 points)	Exceeds Expectations (4 points)	Score
SENTENCE STRUCTURE, DESCRIPTION, AND DIALOGUE	Component is missing or unrelated.	Sentence structure interferes with the flow of the story, and no details are included. Dialogue is not used.	Sentence structure is repetitive and may interfere with the flow of the story. Few details are used. Dialogue is not used appropriately.	Sentences are structured correctly, with some variety. Sentences contribute to the flow of the story and contain some description. Dialogue is used appropriately.	Sentence structure is clear and easy to understand and contributes to the flow of the story. Sentences use engaging description to enhance the story. Dialogue is used appropriately and enhances the story.	
GRAMMAR AND SPELLING	No punctuation marks are used and/ or most words are spelled incorrectly.	Few punctuation marks are used, and many words are spelled incorrectly.	Punctuation is used incorrectly, and there are spelling errors.	Proper punctuation and spelling are used in most places with a few errors.	Writing demonstrates a strong grasp of proper punctuation and spelling, and there are few or no errors.	

TOTAL SCORE: _____

COMMENTS:

APPENDIX C

CONTENT STANDARDS ADDRESSED IN THIS MODULE

NEXT GENERATION SCIENCE STANDARDS

Table C1 (p. 166) lists the science and engineering practices, disciplinary core ideas, and crosscutting concepts this module addresses. The supported performance expectations are as follows:

- 2-PS1-2. Analyze data obtained from testing different materials to determine which materials have the properties that are best suited for an intended purpose.

- 2-PS1-3. Make observations to construct an evidence-based account of how an object made of a small set of pieces can be disassembled and made into a new object.

- 2-LS4-1. Make observations of plants and animals to compare the diversity of life in different habitats.

- 2-ESS1-1. Use information from several sources to provide evidence that Earth events can occur quickly or slowly.

- ETS1-1. Ask questions, make observations, and gather information about a situation people want to change to define a simple problem that can be solved through the development of a new or improved object or tool.

- ETS1-2. Develop a simple sketch, drawing, or physical model to illustrate how the shape of an object helps it function as needed to solve a given problem.

- ETS1-3. Analyze data from tests of two objects designed to solve the same problem to compare the strengths and weaknesses of how each performs.

Table C1. *Next Generation Science Standards (NGSS)*

Science and Engineering Practices

ANALYZING AND INTERPRETING DATA

Analyzing data in K–2 builds on prior experiences and progresses to collecting, recording, and sharing observations.
- Use observations (firsthand or from media) to describe patterns in the natural world in order to answer scientific questions.

DEVELOPING AND USING MODELS

Modeling in K–2 builds on prior experiences and progresses to include using and developing models (i.e., diagram, drawing, physical replica, diorama, dramatization, storyboard) that represent concrete events or design solutions.
- Use a model to represent relationships in the natural world.

PLANNING AND CARRYING OUT INVESTIGATIONS

Planning and carrying out investigations to answer questions or test solutions to problems in K–2 builds on prior experiences and progresses to simple investigations, based on fair tests, which provide data to support explanations or design solutions.
- Make observations (firsthand or from media) to collect data that can be used to make comparisons.

Disciplinary Core Ideas

LS1.C. ORGANIZATION FOR MATTER AND ENERGY FLOW IN ORGANISMS
- All animals need food in order to live and grow. They obtain their food from plants or from other animals. Plants need water and light to live and grow.

ESS3.A. NATURAL RESOURCES
- Living things need water, air, and resources from the land, and they live in places that have the things they need. Humans use natural resources for everything they do.

ESS2.D. WEATHER AND CLIMATE
- Weather is the combination of sunlight, wind, snow or rain, and temperature in a particular region at a particular time. People measure these conditions to describe and record the weather and to notice patterns over time.

Crosscutting Concepts

PATTERNS
- Patterns in the natural and human designed world can be observed, used to describe phenomena, and used as evidence.

SYSTEMS AND SYSTEM MODELS
- Systems in the natural and designed world have parts that work together.

CAUSE AND EFFECT
- Events have causes that generate observable patterns.

Source: NGSS Lead States. 2013. *Next Generation Science Standards: For states, by states.* Washington, DC: National Academies Press. *www.nextgenscience.org/next-generation-science-standards.*

Table C2. Common Core Mathematics and English Language Arts (ELA) Standards

MATHEMATICAL PRACTICES	READING STANDARDS

MATHEMATICAL PRACTICES

- MP1. Make sense of problems and persevere in solving them.
- MP2. Reason abstractly and quantitatively.
- MP3. Construct viable arguments and critique the reasoning of others.
- MP4. Model with mathematics.
- MP5. Use appropriate tools strategically.
- MP6. Attend to precision.
- MP7. Look for and make use of structure.
- MP8. Look for and express regularity in repeated reasoning.

MATHEMATICAL CONTENT

- 2.NBT.A.1. Understand that the three digits of a three-digit number represent amounts of hundreds, tens, and ones; e.g., 706 equals 7 hundreds, 0 tens, and 6 ones.
- 2.NBT.A.2. Count within 1,000; skip-count by 5s, 10s, and 100s.
- 2.NBT.A.3. Read and write numbers to 1,000 using base-ten numerals, number names, and expanded form.

READING STANDARDS

- RI.2.1. Ask and answer such questions as *who, what, where, when, why,* and *how* to demonstrate understanding of key details in a text.
- RI.2.3. Describe the connection between a series of historical events, scientific ideas or concepts, or steps in technical procedures in a text.
- RI.2.7. Explain how specific images (e.g., a diagram showing how a machine works) contribute to and clarify a text.
- RI.2.8. Describe how reasons support specific points the author makes in a text.
- RI.2.9. Compare and contrast the most important points presented by two texts on the same topic.

WRITING STANDARDS

- W.2.1. Write opinion pieces in which they introduce the topic or book they are writing about, state an opinion, supply reasons that support the opinion, use linking words (e.g., *because, and, also*) to connect opinion and reasons, and provide a concluding statement or section.
- W.2.2. Write informative/explanatory texts in which they introduce a topic, use facts and definitions to develop points, and provide a concluding statement or section.
- W.2.6. With guidance and support from adults, use a variety of digital tools to produce and publish writing, including in collaboration with peers.
- W.2.7. Participate in shared research and writing projects (e.g., read a number of books on a single topic to produce a report; record science observations).
- W.2.8. Recall information from experiences or gather information from provided sources to answer a question.

Continued

Table C2. (*continued*)

	SPEAKING AND LISTENING STANDARDS
	• SL.2.1. Participate in collaborative conversations with diverse partners about *grade 2 topics and texts* with peers and adults in small and larger groups. • SL.2.2. Recount or describe key ideas or details from a text read aloud or information presented orally or through other media. • SL.2.3. Ask and answer questions about what a speaker says in order to clarify comprehension, gather additional information, or deepen understanding of a topic or issue.

Source: National Governors Association Center for Best Practices and Council of Chief State School Officers (NGAC and CCSSO). 2010. *Common core state standards.* Washington, DC: NGAC and CCSSO.

Table C3. National Association for the Education of Young Children (NAEYC) Standards

NAEYC Curriculum Content Area for Cognitive Development: Science and Technology
• 2.G.02. Children are provided varied opportunities and materials to learn key content and principles of science.
• 2.G.03. Children are provided varied opportunities and materials that encourage them to use the five senses to observe, explore, and experiment with scientific phenomena.
• 2.G.04. Children are provided varied opportunities to use simple tools to observe objects and scientific phenomena.
• 2.G.05. Children are provided varied opportunities and materials to collect data and to represent and document their findings (e.g., through drawing or graphing).
• 2.G.06. Children are provided varied opportunities and materials that encourage them to think, question, and reason about observed and inferred phenomena.
• 2.G.07. Children are provided varied opportunities and materials that encourage them to discuss scientific concepts in everyday conversation.
• 2.G.08. Children are provided varied opportunities and materials that help them learn and use scientific terminology and vocabulary associated with the content areas.
• 2.H.02. All children have opportunities to access technology (e.g., tape recorders, microscopes, computers) that they can use.
• 2.H.03. Technology is used to extend learning within the classroom and integrate and enrich the curriculum.

Source: National Association for the Education of Young Children (NAEYC). 2005. *NAEYC early childhood program standards and accreditation criteria: The mark of quality in early childhood education.* Washington, DC: NAEYC.

Table C4. 21st Century Skills from the Framework for 21st Century Learning

21st Century Skills	Learning Skills and Technology Tools	Teaching Strategies	Evidence of Success
INTERDISCIPLINARY THEMES	• Economic, Business, and Entrepreneurial Literacy • Health Literacy • Environmental Literacy	• Provide students with the opportunity to investigate natural hazards in the context of the business, economics, and industry of everyday life.	• Students communicate their prior experiences with natural hazards in the context of everyday life.
LEARNING AND INNOVATION SKILLS	• Creativity and Innovation • Critical Thinking and Problem Solving • Communication and Collaboration	• Facilitate creativity and innovation through having students design a natural hazard preparedness plan and public service announcement (PSA). • Facilitate critical thinking and problem solving through use of the engineering design process and having students make observations about a real-world natural hazard occurrence.	• Students demonstrate creativity and innovation, critical thinking, and problem solving as they develop a natural hazard preparedness plan and PSA. • Students work collaboratively and communicate effectively in teams to complete a group project.
INFORMATION, MEDIA, AND TECHNOLOGY SKILLS	• Information Literacy • Media Literacy • Information, Communications, and Technology Literacy	• Engage students in guided practice and scaffolding strategies through the use of developmentally appropriate books, videos, and websites to advance their knowledge.	• Students acquire and use deeper content knowledge as they work to complete natural hazard research and create their preparedness plans and PSAs.
LIFE AND CAREER SKILLS	• Flexibility and Adaptability • Initiative and Self-Direction • Social and Cross-Cultural Skills • Productivity and Accountability • Leadership and Responsibility	• Facilitate student collaborative teamwork to foster life and career skills.	• Throughout this module, students collaborate to conduct research and work on their group project.

Source: Partnership for 21st Century Learning. 2015. Framework for 21st Century Learning. *www.p21.org/our-work/p21-framework.*

Table C5. English Language Development (ELD) Standards

ELD STANDARD 1: SOCIAL AND INSTRUCTIONAL LANGUAGE

English language learners communicate for Social and Instructional purposes within the school setting.

ELD STANDARD 2: THE LANGUAGE OF LANGUAGE ARTS

English language learners communicate information, ideas and concepts necessary for academic success in the content area of Language Arts.

ELD STANDARD 3: THE LANGUAGE OF MATHEMATICS

English language learners communicate information, ideas and concepts necessary for academic success in the content area of Mathematics.

ELD STANDARD 4: THE LANGUAGE OF SCIENCE

English language learners communicate information, ideas and concepts necessary for academic success in the content area of Science.

ELD STANDARD 5: THE LANGUAGE OF SOCIAL STUDIES

English language learners communicate information, ideas and concepts necessary for academic success in the content area of Social Studies.

Source: WIDA. 2012. 2012 amplification of the English language development standards: Kindergarten–grade 12. *https://wida.wisc.edu/teach/standards/eld.*

INDEX

Page numbers printed in **boldface type** indicate tables, figures, or handouts.

INDEX